"A truly impressive work that plugs into an important knowledge gap in the overall global discussion of climate change which so far has been dominated by natural science and hardcore economics. The book is a study of the Pattinavars, a marginalized and traditional marine fishing community in South-eastern India. It focuses on livelihoods and social contexts and argues how concepts of "local adaptation," "vulnerability," and "sustainability" are central to bring into the discussion of climate change as they highlight local coping strategies and community-based traditional knowledge of the natural environment and climate variations. The book is furthermore an important contribution to coastal studies in other parts of the world."

Esther Fihl, Professor Emerita of Cross-Cultural and Regional Studies at the University of Copenhagen, Copenhagen, Denmark

"As a researcher of climate change adaptation measures being undertaken in South Asia, I find great value in this book. We need high quality, thick descriptions of climate change impacts on this region. This analysis underscores the overlapping vulnerabilities that can impact regions as they face multiple waves of shocks. By framing this study in terms of the region's social-ecological system, Dr. Madhanagopal provides a critical account of how our vulnerable regions are adapting to, and falling victim to, the ever-changing nature of climate and the adaptive capacity of human beings."

Christopher Koliba, Ph.D. Professor, Community Development and Applied Economics Department, University of Vermont (UVM), Burlington, Vermont, USA

"Despite all the adaptation and mitigation efforts, climate change continues to be the primary concern, especially for coastal communities in hard-hit regions – those in the frontline. There's much to be learned about how communities cope and deal with disaster risk, and vulnerability associated with climate change, and D. Madhanagopal does a great job in portraying this, with the case of a fishing community in South India, Pattinavars. Through thick description and in-depth analysis, the book reminds us that it is the 'experience near' people like the fishers of Pattinavars who can tell us a great deal about what it means to live in disaster prone areas, and why it is important to 'get the institution right' when dealing with climate change."

Ratana Chuenpagdee, Science Director, TBTI Global Foundation & University Research Professor, Memorial University of Newfoundland, St. John's, Canada

"This insightful and timely work offers a comprehensive analysis of the climate change vulnerabilities and adaptations of Pattinavar marine fishers in South India. Dr. Devendraraj Madhanagopal skillfully illuminates how Indigenous knowledge, local institutions, and gender relations influence survival strategies within this frontline community. This book is essential

reading detailing how climate change is affecting people's livelihoods, and how place-based societies are responding to current challenges."

Brett Clark, Author of *The Tragedy of the Commodity* and *The Robbery of Nature* and Professor of Sociology and Sustainability studies at the University of Utah, Salt Lake City, Utah, USA

"This book reminds us that fishing communities have developed livelihoods in coastal areas for millennia. Their traditional environmental knowledge, accumulated through generations, has helped to overcome natural hazards. Their traditional institutions have also helped in this task. However, transformations induced by climate change make us wonder whether this knowledge is still capable of confronting current challenges or how it needs to be intertwined with scientific knowledge. Likewise, the fit of traditional institutions to these challenges is analysed in this book. This book advances in uncovering the social side of climate change and natural hazards using social science methodologies and qualitative analysis focused on the diversity of small-scale fishing populations. involved, with a particular focus on women. This is an insufficiently developed field in the literature, and this book constitutes an outstanding contribution."

José J. Pascual-Fernández, Professor of Social Anthropology, Director of the Institute of Social Research and Tourism, University of La Laguna, Tenerife, Spain

"In the global discourse of climate change and adaptation, there is a massive need to bring the voices of populations to the margin. They have been experiencing climate-related disasters for generations, and in that, they have developed sophisticated yet pragmatic approaches to deal with the impacts of climate change. Dr. Madhanagopal takes us through the vulnerabilities and sociocultural strengths of the community in Pattinavars in this thought-provoking and well-grounded text. This timely book argues for the need to look at climate change impacts and adaptation beyond scientific lenses and the necessity for understanding social, political, cultural, and economic aspects of the lived experiences of communities. This is a critical read for those who are involved in research and policymaking as well as interested in climate change impact experiences of communities."

Janaka Jayawickrama, Ph.D., University of York, York, United Kingdom

Local Adaptation to Climate Change in South India

This book critically discusses the vulnerabilities and local adaptation actions of the traditional marine fishers of the tsunami-hit coastal regions of South India to climate change and risks, with an emphasis on their local institutions. Thereby, it offers a comprehensive account of the ways in which marine fishers live and respond to climate change.

The Coromandel coastal regions of South India are known for their rich sociocultural history and enormous marine resources, as well as their long history of vulnerability to climate change and disasters, including the 2004 tsunami. By drawing cases from the tsunami-hit fishing villages of this coast, this book demonstrates that indigenous knowledge systems, climate change perceptions, sociocultural norms, and governance systems of the fishers influence and contest the local adaptation responses to climate change. By foregrounding the real picture of vulnerability and adaptation actions of marine fishers in the face of climate change and disasters, this book also challenges the conventional understanding of local institutions and fishers' knowledge systems. Underlining that adaptation to climate change is a sociopolitical process, this book explores the potentials, limits, and complexities of local adaptation actions of marine fishers of this coast and offers novel insights and climate change lessons gleaned from the field to other coasts of India and around the world.

This book will be of great interest to students, scholars, and policymakers in climate change, fisheries, environmental sociology, environmental anthropology, sustainable livelihoods, and natural resource management.

Devendraraj Madhanagopal (Ph.D.) is an Assistant Professor (I) in the School of Sustainability at XIM University (Odisha, India). He holds a Ph.D. in Sociology from the Department of Humanities and Social Sciences, Indian Institute of Technology Bombay (Mumbai, India). He is the recipient of several international travel grants/fellowships. His works appear in *Environment, Development and Sustainability* and *Metropolitics* journals. He is the corresponding editor of the following edited books: (i) *Environment, Climate, and Social Justice: Perspectives and Practices from the Global South* (2022), (ii) *Climate Change and Risk in South and Southeast Asia: Sociopolitical Perspectives*, Routledge, UK (2023), and (iii) *Social Work and Climate Justice: International Perspectives*, Routledge, UK (2023).

Routledge Studies in Hazards, Disaster Risk and Climate Change

Series Editor: Ilan Kelman, Professor of Disasters and Health at the Institute for Risk and Disaster Reduction (IRDR) and the Institute for Global Health (IGH), University College London (UCL)

This series provides a forum for original and vibrant research. It offers contributions from each of these communities as well as innovative titles that examine the links between hazards, disasters and climate change, to bring these schools of thought closer together. This series promotes interdisciplinary scholarly work that is empirically and theoretically informed, with titles reflecting the wealth of research being undertaken in these diverse and exciting fields.

Empowerment and Social Justice in the Wake of Disasters
Occupy Sandy in Rockaway after Hurricane Sandy, USA
Sara Bondesson

Climate Change and Risk in South and Southeast Asia
Sociopolitical Perspectives
Edited by Devendraraj Madhanagopal and Salim Momtaz

Gender-Based Violence and Layered Disasters
Place, Culture and Survival
Nahid Rezwana and Rachel Pain

The Post-Earthquake City
Disaster and Recovery in Christchurch, New Zealand
Paul Cloke, David Conradson, Eric Pawson and Harvey C. Perkins

Local Adaptation to Climate Change in South India
Challenges and the Future in the Tsunami-hit Coastal Regions
Devendraraj Madhanagopal

For more information about this series, please visit: https://www.routledge.com/Routledge-Studies-in-Hazards-Disaster-Risk-and-Climate-Change/book-series/HDC

Local Adaptation to Climate Change in South India

Challenges and the Future in the Tsunami-hit Coastal Regions

Devendraraj Madhanagopal

LONDON AND NEW YORK

First published 2023
by Routledge
4 Park Square, Milton Park, Abingdon, Oxon OX14 4RN

and by Routledge
605 Third Avenue, New York, NY 10158

Routledge is an imprint of the Taylor & Francis Group, an informa business

© 2023 Devendraraj Madhanagopal

The right of Devendraraj Madhanagopal to be identified as author of this work has been asserted in accordance with sections 77 and 78 of the Copyright, Designs and Patents Act 1988.

All rights reserved. No part of this book may be reprinted or reproduced or utilised in any form or by any electronic, mechanical, or other means, now known or hereafter invented, including photocopying and recording, or in any information storage or retrieval system, without permission in writing from the publishers.

Trademark notice: Product or corporate names may be trademarks or registered trademarks, and are used only for identification and explanation without intent to infringe.

British Library Cataloguing-in-Publication Data
A catalogue record for this book is available from the British Library

ISBN: 978-1-032-03511-6 (hbk)
ISBN: 978-1-032-03513-0 (pbk)
ISBN: 978-1-003-18769-1 (ebk)

DOI: 10.4324/9781003187691

Typeset in Times New Roman
by SPi Technologies India Pvt Ltd (Straive)

Contents

List of figures ix
List of tables xi
List of abbreviations and acronyms xii
Acknowledgments xiv

1 Coromandel Coast of South India: centuries-old fishing communities under climate change and disasters   1

2 Conceptual underpinnings and introducing the framework   17

3 Unraveling climate change and disasters: navigating to the field   36

4 Indigenous knowledge systems in confronting climate change: opportunities and constraints   65

5 Living with climate change: vulnerability and adaptation actions   87

6 Local institutions: boon or bane in local adaptation to climate change?   110

7 Fisherwomen and their agencies: scope and challenges in adapting to climate change   140

8 Conclusion and the way forward 162

References 178
Index 218

Figures

1.1	*Akananuru, Neytal* 320	7
1.2	*Akananuru, Neytal* 340	7
1.3	Locating the research region. (a) India (Tamil Nadu is highlighted). (b) Tamil Nadu (Nagapattinam district is highlighted)	9
1.4	Fishing castes in Tamil Nadu – A brief understanding	10
2.1	Adaptation cycle	23
2.2	Adaptation, Institutions, and Livelihoods framework	31
2.3	"Assets/Capitals based Adaptation, Institutions, and Livelihoods (AIL) framework"	34
3.1	The front cover of The Hunger Project report (Perunthottam gram panchayat where Chavadikuppam is located)	44
3.2	Evacuation route planning of Chavadikuppam in disaster situations (a sample)	45
3.3	Tsunami memorial tombs in the study villages	46
3.4	Locating the study villages	48
3.5	Village map of Koozhayar	49
3.6	A small fish market center in Chinnakottaimedu – Also known as Olaikottaimedu (By Red Een Kind as a part of the tsunami relief program)	51
3.7	Fishing vessels and nets	51
3.8	Low-lying areas in Kottaimedu	52
3.9	Fish selling place – Built by IFAD as a part of tsunami rehabilitation measures. Place: Madavamedu	53
3.10	Fishing vessels of Chinnamedu fishers (Both FRP-OBM units and small non-motorized canoes)	54
3.11	Fishing vessels. Place: Chavadikuppam beach	54
4.1	Senior fishermen's perceptions of climate change	74
4.2	The fishing vessels and shrinking coastal spaces. Place: Kottaimedu coast	77
5.1	Fishing vessels of small-scale fishers	91

x *Figures*

5.2	A small-scale fisherman is repairing his fishing net	94
5.3	Fishing vessels	97
5.4	Surukku valai fishing nets	98
5.5	Shrinking coastal spaces	99
6.1	The typical uur panchayat meeting	115
6.2	Panchayat office	118
6.3	Coastal bio-shields	128
6.4	Focus group discussions with fishermen	130
6.5	Focus group discussions with the senior fishermen and local leaders	131
6.6	Coastal sand dunes made by fishers	132

Tables

2.1	Capitals/assets of marine fishers	33
3.1	Vulnerability of coastal Tamil Nadu to climate change and extremes	38
3.2	Nagapattinam district: a saga of climate extreme events	40
3.3	Details of the respondents	58
3.4	Age group and fishing experience of the fisher respondents	59
4.1	A few selected questions/schedules that guided the data collection	67
4.2	A brief illustration of fishermen's local traditional knowledge	70
4.3	Reasons for fish decline and reduction in fish catch: fishermen's perceptions	71
4.4	Comparison of climate perceptions of fishermen with the recent scientific findings	80
7.1	Gendered divisions of small-scale fisheries – focusing on the Pattinavar fishing community	144

Abbreviations and acronyms

AIADMK	All India Anna Dravida Munnetra Kazhagam
BMTPC	Building Material and Technology Promotion Council (Government of India)
CARE	Cooperative for Assistance and Relief Everywhere
CASA	Church's Auxiliary for Social Action
CBA	*Community-Based* Adaptation
CCA	Climate Change Adaptation
CMFRI	Central Marine Fisheries Research Institute
CSTEP	Center for Study of Science, Technology and Policy
DMK	Dravida Munnetra Kazhagam
DRR	Disaster Risk Reduction
ETRP	Emergency Tsunami Reconstruction Project
FAO	Food and Agriculture Organization
FIMSUL	Fisheries Management for Sustainable Livelihoods
FRP	Fibre Reinforced Plastic
GADM	The Database of Global Administrative Areas
GRDRR	Global Facility for Disaster Reduction and Recovery
ICOR	Institute for Community Organisation Research.
ICRAF	World Agroforestry Centre
IFAD	International Fund for Agricultural Development
IK	Indigenous knowledge
INC	Indian National Congress
INCOIS	Indian National Centre for Ocean Information Services
IPCC	Intergovernmental Panel on Climate Change
ISRO	Indian Space Research Organisation
MGNREGA	Mahatma Gandhi National Rural Employment Guarantee Act (MGNREGA)
MoA	Union Ministry of Agriculture
MoEF	Ministry of Environment, Forest and Climate Change (Government of India)
MoSTE	Ministry of State, Technology and Environment (Kathmandu, Nepal)
MSSRF	MS Swaminathan Research Foundation
NABARD	National Bank for Agriculture and Rural Development

NCAER	National Council of Applied Economic Research	
NCW	National Commission for Women	
NDMA	National Disaster Management Authority	
NGO	Non-Governmental Organization	
OBC	Other Backward Classes	
OBM	Out-Board Motor	
OECD	Organisation for Economic Co-operation and Development	
OXFAM	Oxford Committee for Famine Relief	
PRI	Panchayat Raj Institutions	
RBI	Reserve Bank of India	
SC	Scheduled Castes	
SHG	Self-Help Groups	
SIFFS	South Indian Federation of Fishermen Societies	
SNEHA	Social Need Education and Human Awareness	
SoE	State of *Environment* (*SoE*) Report (Government of Tamil Nadu)	
SREX	Special Report on managing the risks of Extreme events and disasters to advance climate change adaptation (Intergovernmental Panel on Climate Change)	
ST	Scheduled Tribes	
TNSAPCC	Tamil Nadu State Action Plan on Climate Change	
TNSCC	Tamil Nadu State Climate Change Cell	
TNSDMA	Tamil Nadu State Disaster Management Authority	
UNDP	United Nations Development Programme	
UNEP	United Nations Environment Programme	
UNFCC	United Nations Framework Convention on Climate Change	
UNISDR	United Nations International Strategy for Disaster Reduction	
VRCC	Vulnerability reduction of coastal communities	
VRIP	Vulnerability-Resilience Indicator Prototype	

Acknowledgments

This book is the outcome of my doctoral research, which I did at the Indian Institute of Technology Bombay (IIT Bombay) in the Department of Humanities and Social Sciences from 2014 to 2019. I acknowledge the research fellowship the University Grants Commission provided me to pursue a Ph.D. at IIT Bombay. I thank the administration of IIT Bombay for providing financial support to conduct the field visits and conference participation, which was much beneficial.

In the first place, I express my huge thanks to Ilan Kelman (University College London). He was generous with his time and provided insightful feedback and constructive suggestions to help me improve my proposal and the outline of this book. I benefited immensely from his in-depth understanding and excellent knowledge of this subject. His motivation, encouragement, and continued inspiration for the development of this book were invaluable. My big thanks to Ilan! Also, I sincerely thank three anonymous reviewers for their insightful comments and valuable suggestions on my book proposal.I thank my Ph.D. supervisor and examiners, Sarmistha Pattanaik, D. Parthasarathy, and K. Narayanan of the Department of Humanities and Social Sciences at IIT Bombay, and Unmesh Patnaik of TISS for their support at various stages of my Ph.D. research. Although I had never met Divya Chandrasekhar of the University of Utah in person, she responded to my e-mails in 2015 and provided her suggestions for advancing this study. I want to thank her for the advice she gave me on identifying the respondents. Ahana Lakshmi shared suggestions that helped narrow down the study regions, and I thank her input. Thanks to Rengasamy (Ret. Professor) and Marirajan for connecting me with several CMFRI and Rameshwaram resource people so I could undertake my initial fieldwork. Similarly, I would like to thank Chris Antony, Berlin Crasian (from the Kanyakumari district), and Vijayan (from Thiruvananthapuram) for giving me knowledge related to the plight of marine fishers. However, I later changed my study area. Thanks to all of them for kindly sharing their time and interest with me in sharing the relevant information.

The practical challenges, vulnerabilities, hardships, unique circumstances, and potentials of maritime fishing communities of the Coromandel Coast of Tamil Nadu were learned through discussions with the local NGO employees.

They were the ones who helped me along and made my fieldwork go smoothly. I would like to express my sincere thanks to the IFAD staff in Nagapattinam district: Bhaskaran, Arul, Senthil, Balamurugan, Ravichandran, and Senthil for sharing the details, including documents, relevant to the fishing villages where I conducted the research. They were kind enough to bring me in touch with several people from the fishing villages, who served as key informants for my fieldwork. I also want to thank V. Selvam (MSSRF) and A. Gurunathan (Dhan Foundation, Madurai) for sharing some valuable information about marine fishers in Tamil Nadu when this study was just starting. Even though I shifted my research location following the pilot survey, I am thankful for their willingness to share information about fishing communities in Tamil Nadu. My sincere thanks to all the participants of this study, including the fisherwomen, fishermen, uur panchayat leaders, other local officials, and panchayat office personnel of the fishing villages, who participated willingly and shared their thoughts and the required details with me. They graciously welcomed me to their fishing villages and households, shared the relevant details, and permitted me to use them for this study. I am grateful to all of them, and I hope this book will contribute to an understanding of their capacity to adapt to any climate crisis.

Several parts of this book were presented in both national and international conferences, including MARE Conference 2017 (University of Amsterdam), 4th International Symposium on the effects of climate change on the world's oceans (Washington, DC), Adaptation Futures 2018 (Cape Town), DSA 2018 (University of Manchester), São Paulo School of Advanced Science on Ocean Interdisciplinary Research and Governance (University of Sao Paulo), Sustainability and Development conference (University of Michigan), Water Security and Climate Change conference (Kenyatta University), Ocean Visions 2019 (Georgia Institute of Technology), The Future Oceans2 IMBeR Open Science Conference 2019 (Brest, France), OceanObs'19 (Honolulu, Hawaii), International Seminar on Poverty, Environment and Sustainable Development Goals in Asia-Pacific (Institute for Social and Economic Change, Bangalore), and Second International Conference on Contemporary Debates in Public Policy and Management (Indian Institute of Management Calcutta). The feedback I received from these academic events on my presentations helped develop this study. I thank the organizers of these academic events and the panels I participated and presented my work. I also thank the following international organizations/institutes that provided travel support and grants for my presentations: Intergovernmental Oceanographic Commission; *The Journal of Development Studies* and DSA, UK; Oceanographic Institute (IOUSP), the Inter-American Institute for Global Change Research (IAI), and the Institute of Advanced Studies (IEA/USP); University of Michigan; exceed-Swindon Project, Germany; Georgia Institute of Technology; The Scientific Committee on Oceanic Research; University Corporation for Atmospheric Research (UCAR); Too Big to Ignore, Canada (declined due to official work).

Since 2017, Esther Fihl of the University of Copenhagen has been kind enough to share with me her extensive knowledge and insights on the fishing communities of Tamil Nadu's Coromandel coast. I thank her for sharing her deep knowledge of coastal Tamil Nadu and for kindly sharing her appreciation for this work. Since I first met Ratana Chuenpagdee (TBTI Global Foundation & Memorial University of Newfoundland) in 2019, she has greatly advocated for this study's progress. I thank her for giving me helpful feedback on this work and for sharing encouragement and appreciation. When I approached Christopher Koliba (University of Vermont) at the beginning of 2019, he provided me with insightful feedback on this study. His critical insights and pertinent questions about this work have been helpful to me in developing the final portion of my Ph.D. dissertation, and I thank him for sharing his time to provide feedback and endorsements. Since I communicated with Brett Clark (University of Utah) in 2019, he has been enthusiastic and positive about this study. He took the time to provide insightful comments on this work and was kind enough to recognize the significance of this work. I thank him for giving his time as well as motivation and friendliness. I am thankful to Janaka Jayawickrama (University of York) for his thoughtful remarks on this book and his encouragement to continue. I thank José J. Pascual-Fernández (University of La Laguna) for providing a valuable appreciation of this book and for sharing great encouragement and appreciation. I thank Patrick J. Christie (University of Washington) for providing me with valuable suggestions during my Ph.D. His sharp comments and critical insights were much helpful in developing the final portion of my Ph.D. dissertation.

During the course of my journey as a Ph.D. student and faculty over the past three years, I met with several academics in person and through e-mail, and they shared their insights and encouragement, which inspired me a lot. I thank Daniel P. Aldrich (Northeastern University), Maarten Bavinck (University of Amsterdam), Dr. Julia Novak Colwell (University of New Hampshire), Arun Agrawal (University of Michigan), Subramanian Karuppiah (University of Amsterdam), Mohammad Mahmudul Islam (Sylhet Agricultural University), D. Nandakumar (Fisheries consultant), Carmen G. Gonzalez (Loyola University Chicago), Mark Axelrod (Michigan State University), and George Adamson (King's College London). I thank all of them for sharing their appreciation and encouragement.

I thank the collaborators of my other recently completed and forthcoming books. I have had the good fortune to work with them for the past three years, and it has been beneficial to me in writing this book. I thank Salim Momtaz (University of Newcastle), André Pelser (University of the Free State), Christopher Todd Beer (Lake Forest College), and Bala Raju Nikku (Thompson Rivers University) for sharing their support.

Substantial portions of Chapter 4 and some parts of Chapter 5 were published in the journals *Environment, Development and Sustainability* (2020, Volume 22, Page numbers: 3461–3489) and *Metropolitics* (https://metropolitics.org/Insecure-Lives-Under-Extreme-Climate-Conditions-Insights-from-a-Fishing-Hamlet.html), I thank the editors of these journals for consenting to use these works in my book. I thank Faye Leerink

(Commissioning Editor) and Prachi Priyanka (Editorial Assistant) for providing excellent and encouraging editorial support. Also, I thank Divya Muthu and Fleming Ambrose (Straive) for their careful copyediting support and Elizabeth Spicer (Senior Production Editor) and her team for their production support of this book.

Perhaps most importantly, I am surrounded by encouraging friends and family. My father, sister, and Priyadarshini have all been quite appreciative of all my endeavors. I am delighted to thank them for their unwavering support and understanding. I thank all my colleagues and administration of XIM University for offering me an atmosphere and time to develop this book amidst teaching schedules.

Devendraraj Madhanagopal
Odisha, India

1 Coromandel Coast of South India

Centuries-old fishing communities under climate change and disasters

Climate change: an unprecedented threat to the world

Climate change includes an increase in the inter-annual mean climate and seasonal variability, changes in the mean climatic conditions, and the increasing frequency and intensity of climate extremes and catastrophic events (Tompkins & Adger, 2004). It brings continuous and unpredictable changes in weather patterns, which have had adverse impacts on multiple sectors. In addition to posing a variety of concerns for human health, climate change also poses dangers to biodiversity by upsetting the equilibrium. Recent shifts in global climate patterns are unprecedented in human history over the previous several thousand years, resulting in catastrophic harm to food security, physical infrastructure, environmental systems, and human systems (IPCC, 2007; Chishakwe et al., 2012; IPCC, 2014a; IPCC, 2014b). There is a growing consensus that global warming and the consequent climate variability have had far-reaching repercussions on coastal zones, the ecosystem, and people's livelihoods (IPCC, 2014a; IPCC, 2014b). Therefore, climate change is an unavoidable reality that demands quick intervention both from the political level and at the community level. The Stern review demands a strong international response and policy actions to face climate change. From an economic standpoint, it cautions that the effects of climate change could stifle growth and development (Stern, 2007). Despite the growing evidence of climate change, the scientific community has seriously underestimated its potential impacts (Thornton et al., 2014). Like this, there are, in the recent past, some scholars questioned the growing fear of climate change and its authenticity. For instance, Hulme (2009) points out the distorted and conflicting messages on climate change and argues that there is no such thing as a stable natural climate. He further reminds us that the scientific communities interpret climate change in a misleading way.

Climate change and the ocean: implications to the fisheries and fishers

In recent decades, research has time and again evidently demonstrated that marine species and the entire coastal systems worldwide are at high risk due to global warming. Many such adverse effects are already visible in many

DOI: 10.4324/9781003187691-1

places, and much more is to come (IPCC, 2007; Perry et al., 2010; IPCC, 2014a; IPCC, 2014b; Pörtner et al., 2014). Climate change impacts mix with the existing stressors of marine ecosystems and intensifies the stress on marine resource-dependent communities (Allison et al., 2005; Perry et al., 2010; Vivekanandan, 2011a; Vivekanandan, 2011b; Zacharia et al., 2016). It exacerbates the threat of human mismanagement in natural ecosystems (Walther et al., 2002; Perry et al., 2010). Climate change and climate variability have a tremendous additional impact on the world's oceans, affecting their production in several aspects. Increasing ocean temperatures and acidity pose serious risks to coastal ecosystems, and the world is becoming increasingly aware of them. It will continue to offer countless challenges to millions and millions of coastal residents in the future. People in undeveloped and underdeveloped countries, particularly those that depend significantly on marine resources for their livelihoods and economies, are more likely to be confronted by the consequences of climate change (Brierley & Kingsford, 2009; Hoegh-Guldberg & Bruno, 2010).

In most cases, marine fishing communities are the frontline victims of climate change and risks. FAO's most recent report shows that 58.5 million people are involved in fisheries and aquaculture. Around 600 million people's livelihoods are completely or partially dependent on these sectors, making fishing one of the most critical natural resources (FAO, 2022a). It plays a crucial role in eradicating hunger and reducing poverty and contributes significantly to food security (Convention on Biological Diversity, 2016). However, at the global level, continuous and systematic exploitation of marine resources by the mechanization of the fishing sector over the past six decades has led to the substantial depletion of fisheries, and such overexploitation trends are intensely posing challenges to sustainability (Pauly et al., 2002; Pauly, 2009; Convention on Biological Diversity, 2016). According to Pauly et al. (2002), considerable focus should be paid to large-scale commercial fishing, which has been the core cause of marine resource depletion, rather than the undefined consequences of environmental change. Nonetheless, Pauly (2009) acknowledges the influence of global warming and rising ocean temperatures, as well as their negative consequences on fish populations, in another work.

The unfair and unjust impacts of climate change

Climate change has a long-term impact on human systems, necessitating research and policymaking through the social sciences. The knowledge and contributions of the social sciences are critical for delving deeply into the vulnerability and adaptation interventions of natural resource-dependent populations. Brulle and Dunlap (2015) underline the importance of key social science research in discussing the various facets of climate change. They advocate the role of sociology and social science in understanding climate change and adaptation. Indicating this, they cite the expanding amount of social science literature on climate change in recent years. However, the sociological literature on climate change has been sparse, and it has not been adequately included in the IPCC reports. Bjurström and Polk (2011) emphasize the IPCC Third Assessment Report and contend that it ignores the role of the social sciences area – economics and environmental studies predominate over social sciences. They contend that the powerful IPCC report is dominated by the natural sciences, notably the earth sciences. They remark this:

> Climate change is addressed as a global environmental problem which is detached from its social contexts.
> (Bjurström & Polk, 2011, p. 15)

There is a famous phrase in climate change discourse: "Climate change affects us all, but not equally." Like this, the renowned Stern review (2007) highlights "the effects of climate change are global, intertemporal and highly inequitable." Though climate change is a worldwide issue, the vulnerabilities of every community or social group are defined by their resources and their ability to claim resource utilization (Kelly & Adger, 2000). The effects of climate change are socially and geographically varied. Poor and marginal populations are particularly vulnerable to climate change since their livelihood systems are already stressed (Adger, 1999; Kelman & West, 2009; Olsson et al., 2014). Climate change places an additional burden on individuals who are already vulnerable in the developing and underdeveloped worlds, forcing them to shoulder the disproportionate impacts of climate change (Adger, 2006). In short, inequality and marginalization are the two most important determinants that influence climate change vulnerability (Ribot, 2011). In this regard, it should be underlined that vulnerability is not just tied to poverty; numerous factors other than poverty influence system/community vulnerability (O'Brien & Leichenko, 2000). Climate change interacts with non-climatic stresses and existing vulnerabilities in local communities, exacerbating inequality and putting further strain on their livelihood systems. As a result, not everyone who is referred to as "vulnerable" is poor (Olsson et al., 2014; Otto et al., 2017). However, it is commonly accepted that climate change has had a significant impact on the well-being of poor resource-dependent populations. In the foreseeable future, agriculture and fisheries will undoubtedly bear the brunt of climate change effects (World Bank, 2013).

Climate change implications and adaptation choices are mostly qualitative and contextual. Hence, there is no way to deal with climate change challenges with only scientific and technological answers. Local adaptation to climate change has garnered a growing consensus among researchers and policymakers in recent years. So, it is important to look at how locals are adapting to climate change in the context of their social, cultural, and political settings. When it comes to evaluating climate change impacts and adaptation alternatives, there are two basic approaches: Top-down and bottom-up (UNFCCC, 2006). Top-down approaches mostly involve quantitative assessments and impact models. Top-down approaches are dominated by climate discussions and energy-based interventions (Dessai & Hulme, 2004; Mitchell & Tanner,

2006). Bottom-up approaches to climate variability (Dubash, 2009) acknowledge local institutions, local coping strategies, and community-based traditional knowledge. It aims to empower the local communities, based on their resources, collective action, and decision-making structures, to reduce their vulnerability to climate change. The goal of community-based climate adaptation projects is to make communities more able to deal with heatwaves, heavy rains, frequent droughts, and unpredictable weather patterns (Mitchell & Tanner, 2006; UNFCCC, 2006). Nonetheless, these methods are not without criticism. Biophysical factors and long-term sustainability are not emphasized in these approaches (Dessai & Hulme, 2004; UNFCCC, 2006).

This book, the outcome of my Ph.D. research, adopts the bottom-up approach to climate change adaptation. It fills the knowledge gaps in social and political dimensions of climate change in South India by focusing on the local adaptation of Pattinavars[1] to climate change. The tsunami-hit coastal regions of Tamil Nadu, which is a part of South India, are the subject of this study. The Pattinavar ocean fishing community in these coastal regions has been impacted by the many consequences of climate change, the long-term ramifications of the 2004 Indian Ocean disasters, and the adverse effects of mechanized fishing. The following portion of this chapter provides an outline of the climate change vulnerability of the coasts and marine fisheries of India, and it sets the research setting – the Nagapattinam district in the Coromandel Coast of Tamil Nadu. After that, it describes the unique significance of conducting this bottom-up climate change research by taking the case of the Pattinavar ocean fishing community. In the end, it pays attention to covering the investigative questions of this book and the layout of the chapters.

Climate change and coastal India: cyclones, floods, rains, erratic temperatures, rising seas, and shrinking coastlines

The coastal populations of South Asia are highly vulnerable to climate change threats due to multiple geographical and socioeconomic factors. Half of the disaster events in South Asia between 1990 and 2008 were due to climate extremes. Sea level rise and coastal inundation have worsened the vulnerability of the millions of coastal populations in South Asia (UNEP, 2009; Ahmed & Suphachalasai, 2014). India, one of the world's most densely populated nations, located in South Asia, is highly vulnerable to climate risks and change. India's vulnerability atlas (BMTPC, 2006) shows that 8.5% of India's land is vulnerable to cyclones, and another 5% of the land is vulnerable to floods. Wheeler (2011) conducts a comprehensive assessment of climate risks and vulnerability for 233 countries worldwide, including India, using three risk indicators: increasing weather-related disasters, sea-level rise, and agricultural productivity loss. This analysis shows that India, for the year 2015, ranks as the third most vulnerable nation among all 233 countries concerning extreme weather events. With about 20.6 million people at risk of sea-level rise, India ranked first among all 233 countries in 2008. Furthermore, by

2050, it is projected that around 37.2 million people will be at risk of sea-level rise. Since the 1950s, a decline in monsoon rainfall patterns and an increase in the frequency of massive rains have been seen across India (World Bank, 2013), leading to a rise in floods and droughts. Recent studies (McGranahan et al., 2007; Wong et al., 2014) show that India, which is one of the most populous countries in the world, has the second largest number of people living in the low-elevation coastal zone. Around 6% of India's population lives in the low-elevation coastal zone, and they are at high risk of cyclones and floods. In their recent review, Nath and Behera (2011) point out that climate change literature has largely been about developed nations. They note the lack of data on the existing and future effects of climate change on India's natural resource-dependent people and call for greater regional research. Amrith's (2018) investigation of climate risks shows that India has been affected by regional and global effects of climate change in a big way and that the expected effects could change India's monsoon patterns in the future.

India has a coastline that stretches for around 7517 km. The coastline of mainland India is approximately 5423 km long, and the coastline of islands is approximately 2094 km long (Sanil Kumar et al., 2006). Infochange (2010) reports that about 25% of India's people live in coastal areas and that about 340 communities depend on the coastal fishing sector for their main source of income (including all the maritime states, union territories, and the Andaman and Nicobar Islands). The Indian marine fishing sector can be categorized into five types: (i) Non-motorized artisanal sector (fishers who use country craft with traditional gear), (ii) motorized sector, (iii) mechanized sector using inboard engines of 50–120 HP, (iv) deep-sea fishing with bigger boats (25 m and above), and (v) motors of 120 HP and higher (Immanuel et al., 2003). According to India's recent marine census, over 8.64 lakh households rely directly on marine fishing for a living. About 90% of the fishers are traditional marine fishers, and about 61% of marine fishing families live below the poverty line (CMFRI, 2012a). Hence, India's marine fishing sector contributes greatly to the country's economic growth and provides food security for millions of underprivileged (World Bank, 2010). The impacts of climate change on the productivity, distribution, and migratory patterns of fish species have increasingly become visible across India. It forces fishermen to travel farther to catch fish in the sea, which reduces their overall net income due to the high cost of sea travel. Due to the erratic and unpredictable changes in the climate, marine fishers are forced to invest more in fishing gear and adopt technologies in fishing (Vivekanandan, 2011a; Vivekanandan, 2011b; Salagrama, 2012; IFAD, 2014). Salagrama (2012) discusses how climate change affects Indian marine fisheries by mixing with the already existing stressors. He identifies that climate change is impacting the following: (i) Access and availability of fish and other coastal resources, (ii) fishing systems, (iii) fishing investments, (iv) quality of life and sea-safety concerns, and (v) traditional knowledge, practices, and governance systems of fishers. Global warming and climate change projections have already had adverse effects on the livelihoods and food security of marine fishers across

India, and it is predicted in a range of erratic and unpredictable environmental conditions (Vivekanandan, 2011).

Why coastal Tamil Nadu? Setting the research region

The poems next page are from *Akananuru*, which provides evidence of fishing practices and fisherfolk's lives in ancient Tamil land. *Akananuru* is one of eight anthologies of Sangam literature. Ancient history in Tamil Nadu, Kerala, and parts of Sri Lanka can be traced back to a time period known as the Sangam period, which approximately lasted from 300 BCE until 300 CE. Sangam literature, which is ancient Tamil literature and the earliest known literature in the southern regions of the Indian Subcontinent, is believed to have been produced during this time period. The origins of Sangam literature and the Sangam era have been the subject of a number of theories. Sangam literature divides the Tamil landscape into five geographical *thinais* (*thinai*, in Tamil), each named after a flower that thrives in that region. The term *"thinai"* is used to describe a region as a whole, encompassing its plants, animals, people, customs, cultures, social order, and social organizations. The five *thinais* are *kurinji* (hilly/mountainous region), *paalai* (desert and parched regions), *mullai* (pastoral tract, woodland region, and neighboring areas), *marutam* (wet/agricultural lands and surrounding regions), and *neithal* (coastal area). A wealth of information on the Paravas (fisherfolk of ancient Tamil land) and their way of life, culture, settlements, fishing vessels, nets, occupations, and sea trade can be found in Sangam literature. Other names for *Akanunuru* are *Nedunthokai*, *Ahappattu*, *Ahananuru*, and *Agananuru*. *Akananuru* is a collection of 400 "*Agam*" (roughly translated as "subjective") poems that depict inner emotions and intimate feelings of love and separation. Although the center of the poem is on introspective themes, they also offer much information about the individuals they describe, including their customs, occupations, and social structures, as well as other cultural insights and evidence regarding the kingdoms and rulers of those eras. This above-noted poem (Figure 1.1), which was written by Madurai Koolavanikan Cheethalai Chatanar, is a classic example. He composed the Tamil-Buddhist epic, *Manimekalai*, and his life period was believed to be probably in the 6th century CE. Hence, it can be assumed that it could have been written sometime around the 6th century CE. The other poem (Figure 1.2) was written by Nakkīraṉār, also spelled as Nakkiranar, and his life period was believed to be probably in the 3rd century BCE (Wikipedia Tamil, n.d.). This poem provides nuanced details of the ocean, sharks, ancient Tamil fishers, and their efficient fishing skills. It should be emphasized in this context that *Akananuru* was written by various poets in three different time layers, from the 1st century BCE to the 5th century CE (Zvelebil, 1974; Narasimhan, 2004; Athiyaman, 2011; Shulman, 2016). A recent study by Balakrishnan (2015) titled *"Sanga elakkiyathil neiyethal nilapanpum nagarikamum"* offers a detailed account of the "Parathava," the ancient Tamil Ocean fishing community (*Neythal thinai*). In this study, Balakrishnan (2015) investigates the

ஓங்கு திரைப் பரப்பின் வாங்கு விசைக் கொளீஇத்,
திமிலோன் தந்த கடுங்கண் வயமீன்,
தழை அணி அல்குல் செல்வத் தங்கையர்,
விழவு அயர் மறுகின் விலை எனப் பகரும்
கானல் அம் சிறுகுடிப் பெருநீர்ச் சேர்ப்ப!.......

Oh lord of the large ocean shore, where
fishermen ride their boats on tall waves
in the wide ocean and pull their nets with
fierce strong fish, which their rich younger
sisters with leaf garments on their loins sell,
calling out prices, on the streets with festivals,
in a beautiful village with groves!.....

Figure 1.1 Akananuru, Neytal 320.
Author: Madurai Koolavanikan Cheethalai Chatanar. Translation from: The Sangam Poems Translated by Vaidehi (Herbert, n.d.).

....திண் திமில்
எல்லுத் தொழின் மடுத்த வல் வினைப் பரதவர்
கூர் வளிக் கடு விசை மண்டலின் பாய்புடன்,
கோட் சுறாக் கிழித்த கொடு முடி நெடுவலை
தண் கடல் அசை வளி எறிதொறும் வினை விட்டு
முன்றில் தாழைத் தூங்கும்,
தெண் கடல் பரப்பின் எம் உறைவு இன் ஊர்க்கே.

The sweet village where we live is on the shores of the

clear ocean, where curved, knotted nets, torn in the

cool ocean by rapidly blowing, harsh winds and

attacking, murderous sharks, sway on thāzhai trees

in front yards where they dry, after the work of the

skilled fishermen end, and on their return from fishing

during the day in their sturdy boats.

Figure 1.2 Akananuru, Neytal 340.
Author: Nakkīraṉār or Nakkiranar. Translation from: The Sangam Poems Translated by Vaidehi (Herbert, n.d.).

rich cultural traditions and lives of *"Parathavas"* by examining the ancient Sangam Tamil literature (including *Akananuru*).

Tamil Nadu, the southernmost state of contemporary India, has a coastline of 1076 km, the second-longest in the nation after Gujarat. There are 14 coastal districts and 608 marine fishing villages in Tamil Nadu. Tamil Nadu's contribution to marine fish production in 2020–21 is estimated at 5.48 lakh tons. A total of 10.48 lakh marine fisherfolk rely on these resources to survive (*Government of Tamil Nadu*, 2022a). Traditional fishers make up 96% of the fishing households in this state. The general economic situation of most marine fishers in Tamil Nadu is poor, with roughly 66% of marine fishing households in the state living in poverty, compared to the national average of 61% (CMFRI, 2012a, 2012b). Therefore, Tamil Nadu's coastlines are renowned for their rich cultural legacy as well as their exposure to climatic extremes and coastal disasters. Between 1900 and 2004, cyclonic storms battered the coasts of this state about 30 times, wreaking havoc. Chennai, Cuddalore, Nagapattinam, Thanjavur, Ramanathapuram, and Kanyakumari are some of the districts that have been impacted (Sundar & Sundaravadivelu, 2005). Southeastern coastal Tamil Nadu is very vulnerable to climate change, according to recent studies (Khan et al., 2012a, 2012b; Ramachandran et al., 2016). During the summer, these areas are particularly susceptible to extreme temperatures (April to May, in common). The rich history of the ocean fishing communities of the early Tamil land dates to the Sangam age. The research is centered on Nagapattinam district, an ancient coastal district in Tamil Nadu, possessing deep-rooted cultural history of the Tamil land. This is part of the Coromandel Coast of Tamil Nadu. Fisheries are one of the primary sources of income for this district. In South India, this coastal stretch is one of the most vulnerable regions to climate risks, change, and coastal disasters.

How does "caste," a social institution, matter in the marine fisheries of India and Tamil Nadu in particular?

For centuries and generations, maritime fishing in the Indian sub-continent has been predominately a caste-based occupation. Historically, the fishers of the coasts of South India have been governed along the lines of caste and gender. Caste – a social institution – continues to impact fishers along India's coasts in numerous dimensions in the contemporary period. Long before the formation of modern government, the Tamil fishing villages were governed by a complex web of socially and culturally based traditions with an unwritten set of customary laws. Even in the contemporary period, these customary norms continue to play an important role in governing their fishing villages, marine fisheries, and sea territories. The maritime governance systems of Tamil Nadu are comprised of a diversified range of formal and non-state (or customary) local institutions that function on caste and gender lines. Traditional local institutions, generally known as fishermen councils, play a major role in the internal governance systems of fishing villages and the

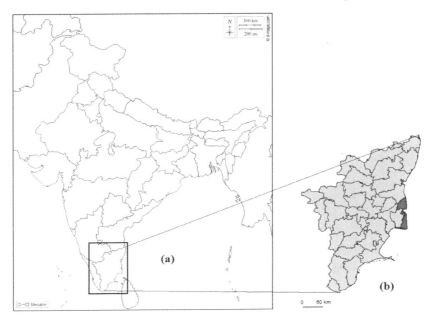

Figure 1.3 Locating the research region. (a) India (Tamil Nadu is highlighted). (b) Tamil Nadu (Nagapattinam district is highlighted). Maps only for illustrative purposes. Not to scale.

Source: (a) https://d-maps.com/; (b) https://gadm.org/.

governance of marine resources without the intervention of external actors (Kurien, 1998). Outsiders find it difficult to enter the marine fishing industry in Tamil Nadu, even in modern times (mainly small-scale fishing). This, however, only applies to small-scale fisheries.

The fishing castes of Tamil Nadu

The three important traditional fishing castes in Tamil Nadu are Pattinavars, Mukkuvars, and Paravas, and they are based in different coasts of the state. The coastline of Tamil Nadu is divided into four zones. (i) Coromandel Coast: The Coromandel Coast stretches for around 357 km from Pulicat to Point Calimere. It encompasses seven coastal districts, including Nagapattinam, the subject of this article's research, and Pondicherry, a Union territory. Pattinavars have mostly lived along this stretch of shore for millennia, primarily engaged in fishing and associated activities. (ii) Palk Bay: It stretches from Point Calimere to Dhanushkodi. Fishing is practiced by a variety of castes in this area. (iii) Gulf of Mannar: The Gulf of Mannar stretches for around 365 km. It is located on the southernmost tip of Tamil Nadu. In this region, fishing is employed mainly by Christian fishermen (Mukkuvar & Paravar). (iv) The West coast is between Kanniyakumari and Neerody. The aforementioned four coastal stretches possess distinct coastal

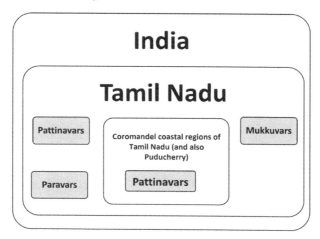

Figure 1.4 Fishing castes in Tamil Nadu – A brief understanding.
Source: Devendraraj Madhanagopal.

ecosystems and fishing systems and distinct cultural identities. The coastal length of the Coromandel Coast is more significant than the other coasts (State of Environment Report for Tamil Nadu, 2016).

The focus of this book is Pattinavars, why?

Pattinavars, a traditional marine fishing community, is historically based in the Coromandel Coastal Stretch of Tamil Nadu and Puducherry[2] for generations and centuries. They inhabit the entire stretch of Tamil Nadu's Coromandel Coast, which lies from Pulicat Lake in the North and Point Calimere to the South. Pattinavars retain their centuries-old rich history in marine fishing and governing the coastal commons (Bharathi, 1999; Bavinck, 2001a, 2001b; Swamy, 2021). As rightly remarked by Gomathy (2006, 2011), for centuries, the Pattinavar people have had a close connection to fishing and the sea, and this connection has naturally shaped the traditional institutions that manage and control the coastal commons. Their traditional institutions are noted for their great internal coherence, self-organization, self-autonomy, and egalitarianism. About 237 fishing villages are on the Coromandel Coast of Tamil Nadu. Of these, 64 marine fishing villages are in the Nagapattinam district, which is 187 km long. Each fishing village along this coastal stretch is networked and operates on the basis of customary governance systems and collective sea-tenure rights. There is a long social and cultural history at the roots of this system (Bavinck, 1996; Gomathy, 2006; Bavinck & Vivekanandan, 2017)[3]. Like Pattinavars, similar patterns of social norms and customary institutional arrangements can be seen among the Mukkuvar fishers.

Bavinck (2011) observes that the differences between Pattinavar's and Mukkuvar's councils lie on the lines of religion. The Mukkuvars are mostly Roman Catholic and, thus, the Church takes primary responsibility for providing institutional infrastructure to the fishers; the parish councils that parish priests head regulate and influence the sociocultural and economic lives of Mukkuvar fishers. With the advancement of scientific developments in recent decades, the fishing methods, infrastructure, and lifestyle of the fishers across the entire coastline of South India, including the Coromandel Coast, have transformed immensely. Their centuries-old norms, values, indigenous knowledge, and customary governance systems, on the other hand, continue to have a huge impact on Pattinavars' lives in a variety of ways. The Pattinavars' customary norms-based internal governance systems are one of India's most well-known examples of self-governance systems. As underscored by Bavinck and Vivekanandan (2017), the fishermen councils (uur panchayats[4]) along the entire coastal stretch of the Coromandel Coast of Tamil Nadu and Puducherry epitomize the classic self-governance systems in South Asia. Alternatively, it is locally known as "*Nattaar panchayats*" and "Uur panchayats" (in Tamil). Uur panchayats can also be considered caste panchayats because the castes of the fishing populations in the villages are mostly homogenous (with a few exemptions). The institutional set-ups and the governance systems of uur panchayats are closely intertwined with the other local institutions. Shared Caste and relations among Pattinavar households have bolstered traditional institutions' authority (Gomathy, 2006, 2011).

Uur panchayats are well-known for their internal coherence, self-autonomy, and self-governance nature, which brings together all the fishing communities along the coastal stretch. It is essential to note that the structure and governance systems of the customary norms of the Pattinavars and their uur panchayats have historically evolved with their culture, traditions, shared values, and intertwined life with the coasts and marine resources. Each fishing hamlet along the coast is administered by its own uur panchayat (Bavinck, 1998; Gomathy, 2006). The uur panchayat[5] of one fishing village is closely bound to the uur panchayat of their adjacent villages, and thereby, they preoccupy strong networks and control over marine throughout the entire stretch of the Coromandel Coast of Tamil Nadu and Puducherry (Bavinck, 1996; Gomathy, 2006; Thamizoli & Prabhakar, 2009; Gomathy, 2011; Bavinck, 2011; Swamy, 2011). The Pattinavars' uur panchayats (fishermen councils) are a typical illustration of how the Indian state allows or overlooks particular social groups (fishers) to govern alternative spaces of political legitimacy, as shown in Esther Fihl's study titled "The legitimacy of South Indian caste councils," which draws on her extensive ethnographic work over the course of four decades on the Coromandel Coast of Tamil Nadu. Although the uur panchayats are labeled as extra-legal bodies with no place in the Indian constitution, fishers do not see or experience them as "alternative spaces" or "extra-legal entities," as she observes. Instead, this is the norm in their eyes, and their conceptions of sovereignty diverge from those of "others" in official India. In addition, the majority of fishers believe that the state is not the "appropriate" place to seek "justice" (Fihl, 2013).

The Coromandel coastal stretch of Tamil Nadu is one of the most potentially vulnerable regions in South India to climate change and coastal disasters. As a part of the Coromandel coast of Tamil Nadu, the Nagapattinam district has been regularly hit by various extreme weather and climate-related events as well as coastal disasters. On December 26, 2004, the fishing villages of the Southern Coromandel coast were devastatingly hit by the tsunami waves, damaging the human, physical, and coastal systems along the entire coastal stretch. Substantial knowledge is available discussing the effects of the Tsunami disaster on Tamil Nadu, including the aftermath, recovery, and long-term consequences for the state's fishers (Srinivasan & Nagarajan, 2005; Prater et al., 2006; Rodriguez et al., 2008; Kuppuswamy & Rajarathnam, 2009; Hastrup, 2011; Kumar, 2017). Like this, a few works discuss the early warning systems of the 2004 tsunami (2004 Indian Ocean Tsunami) disaster, and the impacts of the tsunami on international tourists (Kelman, 2006a; Kelman et al., 2008). A few noted studies exclusively focus on the Nagapattinam district, and they offer a wealth of information and insights on tsunami recovery, rehabilitation of the state and humanitarian agencies, the roles of local actors and institutions in recovery, as well as its complexities, including elite capture and gender inequities (Chandrasekhar, 2010; Hastrup, 2011; Swamy, 2011; Gandhi, 2011). The tsunami disaster resulted in widespread poverty, unemployment, and insecurity in the lives of Pattinavars for at least a decade. The Pattinavar marine fishing communities in this region are still dealing with the long-term effects of the tsunami in numerous ways that are complex and multifaceted.

Investigative questions of this book

The study is based on the selected coastal fishing villages of the Nagapattinam district, which act as a foundation for this book. I chose this region as the focus of this study for three key reasons. (i) This coastal stretch is highly vulnerable to climate change and associated risks, not just in India but also in South Asia.[6] (ii) This was the worst-hit district by the 2004 Indian Ocean Tsunami in the entire Indian subcontinent. The long-term repercussions of the tsunami disaster continue to influence and affect the livelihood stress of Pattinavar fishers, who have predominately resided in these regions for ages. Climate change and the 2004 tsunami's effects act as a double burden to these fishers. (iii) The customary norms and self-governance systems of the Pattinavar fishing community of these coastal regions have a definite social and anthropological significance, and they can offer novel insights to other coastal regions of India and around the world. In the Indian subcontinent, the scope and limits of the traditional ocean fishing community's customary norms and governance systems in adapting to climate change have received little attention. This book addresses these knowledge gaps and provides deep insights into the climate change vulnerability and local adaptation of Pattinavar marine fishers by examining their deep-rooted customary norms and local institutions. This type of literature is scarce in South Asia, making this work immensely significant. The scope of this research includes two distinct themes that are important to the fundamental objectives: (i) Pattinavar indigenous knowledge and (ii) women's agencies in climate change adaptation. Given this backdrop, this book aims to interrogate the following questions:

- How do fishermen perceive indigenous knowledge and climate change, and what are the contemporary importance and relevance of indigenous knowledge systems in responding to climate change?
- How does climate change affect marine fishers and what are their local adaptation responses?
- What are the roles, governance systems, scope, and limits of fishers' local institutions in sustaining adaptation efforts to climate change?
- What are the challenges that newly formed women's agencies face in responding to climate change and adaptation, and how are influential local institutions responding to it?

In the climate change literature focusing on India, there has been little discussion of slow-onset events such as erosion and its effects on traditional ocean fishing communities. This book presents a detailed account of capturing the local perspectives of fishers from tsunami-hit coastal districts on erosion caused by climate change and dangers, addressing one of the few unexplored topics in climate change. It presents a conceptual framework for discussing the local adaptation of marine fishing communities to climate change adaptation, that could be applied to discussing climate change adaptation in an academic discourse focusing on other parts of India as well as South Asia. Overall, this book integrates knowledge from academic, practical, local, and international domains, and thereby, it explores the local adaptation of marine fishing communities to climate change. The insights and discussions of this book can be scaled up to examine the complexities of all traditional ocean fishing communities in Tamil Nadu and India in responding to climate change. It is equally pertinent to note that this research context needs to be considered while scaling up to other settings. The internal governance systems of coastal fishing communities in South India possess certain notable similarities and dissimilarities with other regions. As a result, the book's reviews, conceptual debates, and empirical findings contribute significantly to our knowledge of the social and political elements of climate change among resource-dependent traditional communities in South India, particularly Tamil Nadu. The novel conceptual and empirical insights and recommendations of this book can be highly beneficial to climate change researchers, policymakers, and professionals working in similar sectors.

Book plan

This book is organized sequentially to address the research questions posed. In total, it consists of eight chapters, including the introductory chapter and conclusion. Chapter 2, titled "Conceptual underpinnings and introducing the framework," reviews the conceptual underpinnings of this research and clarifies the conceptual definitions adopted to lead this research. It discusses the Sustainable Livelihoods framework and the Adaptation, Institutions, and Livelihoods (AIL) framework and highlights its influence on fisheries and climate change adaptation studies. It then proposes a unified conceptual framework by amalgamating both frameworks to guide this study. This framework illustrates the capital, governance systems, and adaptation processes of the traditional marine fishing community to climate change.

Chapter 3, titled "Unraveling climate change and disasters: navigating to the field," unravels the climate change complexities along the Coromandel Coast of South India and argues how marine fishers along this coastal stretch are exposed to multiple vulnerabilities, including climate change, climate extremes, and coastal disasters. It highlights the long-standing effects of the 2004 Indian Ocean Tsunami disaster on fishers and explains the ways in which it complicates the weather patterns. Thereby, it provides a double burden to the fishers across this region. The chapter then pays attention to how I selectively identified the study villages along Tamil Nadu's Coromandel Coast after having conducted a rigorous pilot survey. To this end, the chapter elaborates on the reasons behind choosing these villages to conduct this study. The geographical location of the fishing villages and socioeconomic features of these fishing communities will be described in depth in a large chunk of the chapter. The chapter also illustrates how qualitative research methodologies were well suited to satisfy the objectives of this study and how the study was carried out in the field.

Chapter 4, titled "Indigenous knowledge systems in confronting climate change: opportunities and constraints," explores the contemporary importance of indigenous knowledge systems of Pattinavar marine fishers in climate change. By examining indigenous knowledge and climate perceptions of fishers, the chapter discusses the opportunities of how indigenous knowledge systems of place-based communities can complement and address the knowledge gaps that scientific knowledge possesses in investigating climate change and its impacts. It also cautions researchers and policymakers about the problems that persist in overrating the real value of indigenous knowledge systems of place-based communities in confronting climate change. While highlighting the empirical insights, the chapter critically puts forth that climate change impacts are mostly unpredictable, particularly in the coastal regions with a recent history of tsunami disasters.

Chapter 5, titled " Living with climate change: vulnerability and adaptation actions," covers the impacts of climate change on fishers and their local coping and adaptation actions to confront it. Climate change research indicates that economic marginalization exacerbates the vulnerability of marginalized groups. By backing this up, the chapter proposes an alternative view: weak and inadequate social networks and political capital make traditional marine fishers more vulnerable to climate change and makes it harder for them to adapt. The chapter contends that the following sets of factors influence the local adaptation of marine fishers to climate change, showing both conceptual and empirical insights: institutions, social networks, political capital, and population density. It concludes that these elements determine the winners and losers among fishers and fishing villages as they adapt to climate change.

Chapter 6, titled "Local institutions: boon or bane in local adaptation to climate change?," provides valuable insights into the knowledge of local institutions in a traditional coastal fishing community in terms of climate change adaptation. It outlines the range of informal, semi-formal, and formal local institutions of the fishing villages across the coastal stretch of the Coromandel coast of Tamil Nadu. It then describes its roles, functions, and governance systems in the fishing villages. The chapter then critically examines local institutions based on the power and networking capabilities of their primary actors, and discusses how this relates to climate change adaptation. Literature on climate change and resilient communities often portrays local institutions in a positive light, particularly in terms of aiding adaptation. Except for a few notable exceptions, there is a dearth of specialized literature on the topic of local institutional responses to climate change in South Asia. Besides this, the limits of local institutions in climate change adaptation have been largely disregarded in the available literature. In reality, all stakeholders in local institutions are not fairly represented, which is particularly problematic in adapting to climate change. In addition, local institutions, and their capacity to organize the social capital of the community have been exaggerated. This book, which focuses on Pattinavar marine fishers, contends that local institutions' social capital (including networking capability) is successful, subject to its ability to couple with the political and economic capital of the community. On the other hand, in climate change adaptation, value systems embedded in customary governance systems of local institutions have received less attention. The chapter fills in the gaps that have been pointed out by giving selected cases and highlighting local voices, pointing out the possibilities, challenges, and chances that lie ahead for local institutions, and giving interesting insights into how they can help adapt to climate change.

Chapter 7, "Fisherwomen and their agencies: scope and challenges in adapting to climate change," has considerable continuity with Chapter 6. The chapter looks at how women's self-help groups (a relatively new semi-formal local institution) might assist fishing households and communities adapt to climate change. The chapter examines male-dominated local institutions via the views of women's self-help group members, senior fisherwomen, local leaders, and fishermen. It looks at the setbacks of the functions of male-dominated local institutions of the fishing villages in responding to climate change, from the immediate effects of extreme climate conditions to the long-term impacts of climate change on coastal erosion. It then critically discusses how the sociocultural norms of the male-centric local institutions impacted the growth of women's self-help groups, thereby weakening women's agency in responding to climate change and risks. The chapter underscores the need to build local institutions that give fair and equitable representation to all stakeholders, especially women's groups, to adapt to climate change.

Chapter 8, titled "Conclusion and the way forward," begins by examining the research objectives and summarizing the significant findings (conceptual and empirical). It highlights the contributions of this book and discusses the ways in which it brings unique insights to global debates on climate change knowledge and practices of traditional marine fishing communities (including other place-based societies). It discusses the current knowledge gaps, with a particular emphasis on the coastlines of India and how they may be addressed in the future. It ends by pointing out what this book offers and how it benefits coastal studies in other regions of India and the world.

Notes

1 Pattinavar can alternatively be written as "Pattanavars." Throughout this book, I employ the word "Pattinavar" to ensure consistency.
2 A state and a union territory located in the southern region of India, respectively. For an elaborate understanding of the castes and tribes of South India, refer to Thurston and Rangachari (1909).
3 One of the notable academics who studied deeply the uur panchayats and legal governance structures of the Tamil Nadu marine fishing communities over the past few decades is Maarten Bavinck. In this book, I refer to his writings extensively while discussing uur panchayats. Nevertheless, it is essential to mention that his writings on Tamil marine fishing communities deal almost exclusively with marine resources and governance systems, but not with climate change.
4 This book uses the terms "customary village councils," "traditional panchayats," "community panchayats," "customary local institutions," and "fishermen councils" to refer to "uur panchayats." In some instances, the terms are used interchangeably, though these terms possess different meanings based on contexts. In this book, uur panchayats/fishermen councils refer to non-state, informal, and community entities. To minimize conceptual and terminological complexity, this book largely uses the terms "uur panchayats" and "Pattinavar uur panchayats" to refer to the fishermen councils of Pattinavars since they convey a straightforward meaning. The Tamil words "uur panchayats" and "Pattinavar uur panchayats," however, have not been italicized to ensure read with a good flow. In some instances, depending on the context, this book also refers to uur panchayats as traditional village councils and fishermen councils.
5 The following chapters of this book provide in-depth discussions of the uur panchayats' activities, role-making, and governance systems. The purpose of this chapter is to give readers an overview of Pattinavar fishing communities, their significance, and unique fisheries governance systems, as well as to explain why focusing on Pattinavars may help us better understand the social and political dimensions of climate change.
6 Chapter 3 of this book carefully outlines the vulnerability of these locations to climate change and tsunami disasters using scientific evidence.

2 Conceptual underpinnings and introducing the framework

How are climate change, climate variability, and extreme events viewed in this book?

Climate change is a multi-scale global problem (Adger, 2006). The scientific community attributes both natural and anthropogenic processes are the factors of climate change. Climate change results in the complex interactions of both natural and anthropogenic processes and has diverse impacts on the natural and human systems (IPCC, 2014a).

> Climate change refers to a change in the state of the climate that can be identified (e.g., by using statistical tests) by changes in the mean and/or the variability of its properties and that persists for an extended period, typically decades or longer. Climate change may be due to natural internal processes or external forcings such as modulations of the solar cycles, volcanic eruptions, and persistent anthropogenic changes in the composition of the atmosphere or land use.
>
> (IPCC, 2014a)
>
> Climate variability refers to variations in the mean state and other statistics (such as standard deviations, the occurrence of extremes, etc.) of the climate on all spatial and temporal scales beyond that of individual weather events. Variability may be due to natural internal processes within the climate system (internal variability) or to variations in natural or anthropogenic external forcing (external variability).
>
> (IPCC, 2014a)

18 *Conceptual underpinnings and introducing the framework*

Climate change is a long-term effect. Climate change and variability can both occur at the same time. Both climate change and climatic variability increase the intensity and frequency of climate extreme events, a study reveals (IPCC, 2012). Climate change is evident, according to the IPCC report "Summary for policymakers" (IPCC, 2014b). Recent climate change consequences have been felt across oceans and continents. However, the hazards of climate change and extremes are unevenly distributed, and so the effects of climate change on biological and human systems are grossly underestimated. The IPCC consistently underlines in its reports that coasts and coastal systems around the world are at high risk of climate change (IPCC, 2014a, 2014b). The Special Report on Managing the Risks of Extreme Events and Disasters to Advance Climate Change Adaptation (SREX) (IPCC, 2012) presents a holistic review of extreme weather and climate events and disasters – their definitions, classifications, and recent case studies. It recognizes that climate change and extremes will cause unprecedented damage to nature and human systems. Lavell et al. (2012), in the recent IPCC report, define extreme events as

> Extreme events comprise a facet of climate variability under stable or changing climate conditions. They are defined as the occurrence of a value of a weather or climate variable above (or below) a threshold value near the upper (or lower) ends ('tails') of the range of observed values of the variable.
>
> (Lavell et al., 2012, p. 30)

In the IPCC report, Seneviratne et al. (2012) warn that multi-decadal changes in the climate are the main cause of the increase in weather and climatic events. Heavy rainfalls, floods, tropical cyclones, and precipitation will rise globally in the 21st century. However, the authors of this paper observed that not all extreme weather and climate occurrences could have a significant influence on the community. Using temporal frames, they differentiated between extreme weather events and climate events. They employed the word "climate extremes" for the sake of simplicity, as it refers to both extreme weather and climate events. Following this strategy, this book loosely used the phrase "climate extremes," which refers to all extreme weather and climatic events in the field settings. The IPCC report "Summary for policymakers" provide the conceptual clarification on the term "impacts." It explains:

> Effects on natural and human systems. In this report, the term impacts is used primarily to refer to the effects on natural and human systems of extreme weather and climate events and of climate change. Impacts generally refer to effects on lives, livelihoods, health, ecosystems, economies, societies, cultures, services, and infrastructure due to the interaction of climate changes or hazardous climate events occurring within a specific time period and the vulnerability of an exposed society or system. Impacts are also referred to as consequences and outcomes. The impacts

of climate change on geophysical systems, including floods, droughts, and sea level rise, are a subset of impacts called physical impacts.

(IPCC, 2014b, p. 5)

Climate change is more likely to hurt poor and disadvantaged communities in underdeveloped and developing nations (Thornton et al., 2014). Climate change impacts linked with variability and extremes, as observed by Lavell et al. (2012), provide additional difficulties to the vulnerability and risks of many communities, including disadvantaged sections, thereby increasing disaster risk. This may result in the state failing to achieve its disaster risk management goals. Besides, as succinctly pointed out by Cutter et al. (2012), the impacts of disasters associated with climate events are most acutely experienced locally. Hence, there is a strong need to focus on local vulnerability and localized impacts of climate change. This book employs the above-said definition to understand "climate change impacts." Throughout this book, the term "impacts" refers to the broad impacts of climate change associated with variability and "extreme events" on the fishers' lives, livelihoods, economies, and well-being. The threats imposed by climate change on poor and marginal communities are severe and multi-dimensional. In this context, it is pertinent to remember that all the vulnerable groups to climate change are not necessarily poor. Climate change impacts interact with non-climatic stressors and the existing vulnerabilities of the local communities and exacerbate their inequalities, making them further vulnerable (Adger, 1999; Otto et al., 2017).

Vulnerability, resilience, and adaptation to climate change

Vulnerability to climate change

Globally, vulnerability and social vulnerability to climate change and environmental hazards have been studied and discussed for decades (Blaikie et al., 1994; Cutter et al., 2003; Smit & Wandel, 2006; Patnaik & Narayanan, 2009; Kelman, 2011a; Maiti et al., 2015; Kelman et al., 2016; Kelman, 2020). Cutter et al. (2003) lay forth the three central tenets of vulnerability research. They are as follows: (i) the identification of conditions that render individuals or locations susceptible to extreme disasters, (ii) the presumption of vulnerability is a social condition, and (iii) the integration of potential exposures and societal resilience, with a specific focus on specific places or regions. Füssel and Klein (2006) explored the concept of "vulnerability" in the context of climate change and assessed how this term had been discussed. Their assessment shows that conceptual understandings of vulnerability in the context of climate change have evolved and grown through time. The concept of vulnerability is widely used. However, no commonly accepted definition exists (Downing & Patwardhan, 2004). Initial conceptual debates on vulnerability focused mostly on hazard research (Blaikie et al., 1994). The concept of "vulnerability" was widely studied across several disciplines in the 1990s, particularly in climate change (Janssen et al., 2006). This concept has mostly

20 *Conceptual underpinnings and introducing the framework*

been discussed in disasters, hazard risks, hazard mitigation studies, and climate change risks (Cutter, 1996; Adger, 1999; Kelman et al., 2015). In short, vulnerability can be defined as "the potential for loss" (Cutter, 1996). To understand and conceptualize, this study refers to the definition given by the IPCC. "Vulnerability is the propensity or predisposition to be adversely affected. Vulnerability encompasses a variety of concepts and elements including sensitivity or susceptibility to harm and lack of capacity to cope and adapt" (IPCC, 2014b, p. 5).

The multiple dimensions of vulnerability include exposure, sensitivity, and adaptive capacity. To define vulnerability, measurements of both environmental and social systems are required. Social vulnerability varies over space and through time (Cutter & Finch, 2008). Since the 1990s, the vulnerability concept has been widely used in livelihood studies. In this context, vulnerability refers to the condition that makes individuals/households unable to sustain their livelihood. To conceptualize social vulnerability, this study refers to the work of Kelly and Adger (2000):

> The capacity of individuals and social groupings to respond to – that is, to cope with, recover from or adapt to – any external stress placed on their livelihoods and well-being, focusing on socioeconomic and institutional constraints that limit the ability to respond effectively.
> (Kelly & Adger, 2000, p. 347)

The social, political, and economic vulnerabilities of society shape the effects of climate extremes. As a result, vulnerability analysis is critical in lowering the vulnerabilities of climate hazards and change. Ribot et al. (1996) remind us that the regional understanding of climate change is key to planning its effects and implications. In another work, Ribot (2010) emphasizes that the central element of climate change adaptation is vulnerability reduction. Climate change literature analysis on India over the last three decades reveals that much focus was given to scientific studies, but not bottom-up studies on understanding and assessing the impacts, vulnerability, and adaptation options of climate change.

Resilience to climate change

Holling introduced the resilience concept from ecology (Janssen et al., 2006). As a result of Holling's key 1973 paper, *Resilience and Stability of Ecological Systems*, the notion of resilience has extended across a variety of fields. After the 1990s, the idea of "resilience" started to be talked about by experts, especially when it came to social resilience and socioecological systems (Adger, 2000; Berkes et al., 2000). Adger (2000) defines social resilience as "The ability of the communities to withstand external shocks (e.g., environmental risks) and adapt them." Keck and Sakdapolrak's (2013) recent review of the growing body of literature on social resilience categorizes the three capacities of social resilience: (i) Coping capacities: to cope and overcome immediate adversities (re-active – short term), (ii) adaptive capacities: to learn from experiences and adjust to facing new challenges in the future (pro-active – long-term), and (iii) transformative capacities. Voss (2008) and Lorenz (2010) use the term "participative capacity" to denote the transformative capacities of the stakeholders to assess the assets and craft their institutions to promote individual and social robustness to respond to present and future risks (Voss, 2008; Lorenz, 2010; Keck & Sakdapolrak, 2013). Lorenz (2010) points out that participative capacity is mostly neglected in social resilience research. It can be described as the measure of the ability of social systems to self-organize.

Adaptation to climate change

Three reasons force societies to adapt to environmental change. (i) Climate change projections are evident, and the problems posed by environmental change are more apparent and predictable than ever. (ii) Climate change projections and environmental risks may not be quantifiable in notable instances. However, the consequences of the dangers are severe. Adaptation to climate change projections necessitates the transformations of the systems. It also necessitates the coordinated efforts of various actors. In response to a variety of pressures and stimuli, adaptations may falter. (iii) People who are vulnerable and at risk are being asked to adapt to climate change more and more. This is because climate change seriously affects human systems (Nelson et al., 2007). Abeygunawardena et al. (2009) point out that Asia's poor are particularly vulnerable to climate change and extremes, emphasizing the necessity of better governance and access to assets and services for the poor in adapting to climate change. They indicate a few areas where effort should be focused in response to poverty and climate change, as well as the elimination of maladaptation. They are as follows: (i) Better governance mechanisms, (ii) bringing climate change issues and adaptation strategies into the mainstream, (iii) promoting bottom-up approaches and focusing on local institutions and local knowledge, (iv) community empowerment, (v) vulnerability assessments, (vi) "information" accessibility, (vii) impacts integration, and (viii) increasing livelihood resilience. According to Wolf (2011), climate change adaptation is intrinsically a social process that is dependent on the social and cultural characteristics of the stakeholders. Ford and King (2015) list six important factors for adaptation, which are as follows: (i) Political leadership, (ii) institutional organization, (iii) adaptation decision-making and stakeholder engagement, (iv) availability of usable science, (v) funding for adaptation, and (vi) public support for adaptation. These factors may reinforce each other and influence the adaptation readiness of the stakeholders. Adaptation readiness and allocation of funds for adaptation are highly political.

Vulnerability, resilience, adaptation, and adaptive capacity to climate change have all been studied extensively during the past three decades of research. Janssen et al. (2006) present significant seminal papers on three knowledge domains: vulnerability, resilience, and adaptation. There are as follows: Smithers & Smit (1997); Tol et al. (1998); Smit et al. (2000); McCarthy et al. (2001). The IPCC reports are primarily used to define adaptation and related concepts in this book. The IPCC has regularly updated its definition of coping and adaptation in light of the growing corpus of climate change research. The following are the definitions provided in the recent IPCC report.

> Adaptation: "In human systems, the process of adjustment to actual or expected climate and its effects, in order to moderate harm or exploit beneficial opportunities. In natural systems, the process of adjustment to actual climate and its effects; human intervention may facilitate adjustment to expected climate and its effects."
>
> (IPCC, 2022)

> Coping: "It is the use of available skills, resources and opportunities to address, manage and overcome adverse conditions, with the aim of achieving basic functioning of people, institutions, organizations and systems in the short to medium term."
>
> (IPCC, 2022)

There are different ways to adapt to climate change. Anticipatory adaptation, autonomous adaptation, planned adaptation, private adaptation, public adaptation, and reactive adaptation are some of them (Smit et al., 2000; Smit & Pilifosova, 2001; IPCC, 2007 for more discussions on the types of adaptation). To handle the severe demands of climate change and variability, autonomous adaptation is insufficient. Given the potential of climate change problems, planned anticipatory adaptation actions are essential in reducing vulnerability and utilizing climate change opportunities (Smit & Pilifosova, 2001; IPCC, 2007). Given the importance of focusing on disaster risk reduction, this book also sheds light on coping and disaster risk reduction measures of the fishers. As Birkmann et al. (2009) clarified, the field of disaster risk reduction uses the word "coping" to describe how people react, respond, and make decisions to deal with the effects of hazards. In climate change literature, on the other hand, the term "adaptation" is employed. This study utilizes the IPCC's fifth assessment report to define disaster risk reduction.

> Disaster risk reduction denotes both a policy goal or objective and the strategic and instrumental measures employed for anticipating future disaster risk, reducing existing exposure, hazard, or vulnerability, and improving resilience. This includes lessening the vulnerability of people, livelihoods, and assets and ensuring the appropriate sustainable management of land, water, and other components of the environment.
>
> (Lavell et al., 2014, p. 34)

A relatively recent approach to adaptation is a community-based adaptation (CBA) (Huq & Reid, 2007; Reid et al., 2009). This approach is particularly relevant to this book as it focuses on local institutions of marine fishers and

Conceptual underpinnings and introducing the framework 23

adaptation to climate change at the local level. To conceptualize CBA to climate change, this book draws from Reid et al. (2009): "Community-based adaptation to climate change is a community-led process based on communities' priorities, needs, knowledge, and capacities, which should empower people to plan for and cope with the impacts of climate change." CBA focuses on local needs and the contextual nature of climate change vulnerability. Ayers and Forsyth (2009) highlight that CBA approaches are primarily participatory, and it goes in hand with the local development and building of community resilience. It emphasizes bottom-up approaches to identifying the vulnerability and facilitating adaptation actions.

A few important questions on adaptation

To build the adaptation framework and understand the adaptation cycle, Wheaton and Maciver (1999) design four key questions: (i) Who or what adapts?, (ii) What do they adapt to?, (iii) How do they adapt?, and (iv) What and how are resources used? Smit et al. (1999) devise three critical questions: (i) Adapt to what?, (ii) Who or what adapts?, and (iii) How does adaptation occur? (see Figure 2.1). Pelling (2011) expands on these conceptual discussions of climate adaptation by posing a new question: What are the limitations of adaptation? These questions serve as the driving force for framing the objectives of this study, focusing on marine fisheries settings.

Power and social relations play central roles in shaping adaptation actions and influencing the adaptive capacity of the community. The "power" of the crucial actor/stakeholder influences the support for adaptation actions in a

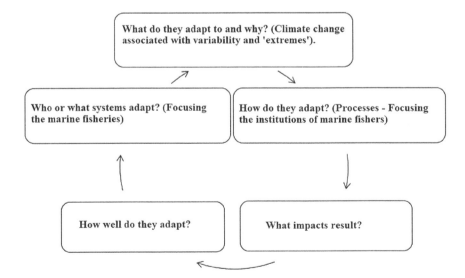

Figure 2.1 Adaptation cycle.
Source: Adapted from Wheaton and Maciver, 1999.

significant way (Pelling, 2011; Pelling & Manuel-Navarrete, 2011). Therefore, investigating how power influences the marine fisherfolk and their local institutions can provide novel insights into the potentials and limits of their coping and adaptation measures in responding to climate change. In this context, it is important to highlight that adaptation is also not free from critiques. Ribot (2011) critiques the term "adaptation" and cautions that it may lead to a Social-Darwinist worldview, implying that those who cannot survive and adapt to threats are unfit. The term "adaptation" should inspire society to help underprivileged people overcome their challenges. The moral obligation of society is to safeguard its citizens, regardless of their backgrounds. On the other hand, the term "adaptation" implies that society diminishes the social, economic, and political contributions of excluded groups. He adds that the adaptation discourse forces research groups to think of answers without addressing the reasons for vulnerabilities, such as why some people are weak and who is responsible for such vulnerabilities, by debating the consequences of vulnerability and adaptation (Ribot, 2010; Ribot, 2011).

Adaptive capacity to climate change

In simplest terms, as defined by Adger et al. (2007), "adaptive capacity is the ability to adapt to shocks and risks. It includes adjustments in both behaviors and resources and technological options. Adaptive capacity is uneven across and within the societies."

According to Smit & Pilifosova (2001), the socioeconomic characteristics of the community determine its adaptive capacity. Whereas Yohe and Tol (2002) note the range of sectors, systems, and factors that determine the adaptive capacity of the community to climate change. They are (i) multiple technological options for adaptation, (ii) availability of adequate resources and equal distributions across the society, (iii) strong and effective institutional set-ups for decision-making processes and governance, (iv) the stock of human and social capital, (v) effective risk-spreading processes, (vi) effective and credible decision-making authorities who can be able to manage and check the details of information, and (vii) Public perceptions of risks. Social capital, institutions, trust, and reciprocity significantly influence the adaptive capacity of the community (Adger, 2003; Pelling & High, 2005). To conceptualize adaptive capacity, this book employs the definition provided by the recent IPCC report: "Adaptive capacity is the ability of systems, institutions, humans, and other organisms to adjust to potential damage, to take advantage of opportunities, or to respond to consequences." However, all adaptation actions are not necessary to promote the adaptive capacity of the system or stakeholders. There are significant limits and barriers to implementing adaptation actions, including (i) physical and ecological limits, (ii) Technological limits, (iii) financial barriers, (iv) information and cognitive barriers, and (v) social and cultural barriers (Adger et al., 2007, pp. 733–737). This study draws understanding from the limitations of adaptation categorized by Adger et al. (2007).

Social capital and social networks

According to Bebbington (1999), capitals are not a resource for people to develop their livelihoods; they give meaning and the power to be and act. This research investigates how the five capitals – Natural, Physical, Economic (Financial), Human, and Social – of marine fishers mediate with their local institutions in vulnerability and adaptation actions to climate change. Among the five capitals/assets, this study focuses on "social capital" because it is particularly relevant to local-level institutions. Social capital is a popular concept in development literature and holds a dominant position in the social sciences. Social capital, according to Woolcock and Narayan (2000), provides the buffer and safety nets for responding to threats, such as poverty or vulnerability, in a society. "Social capitals are the norms and networks that enable people to collaborate" (Woolcock & Narayan, 2000, p. 226). This study indicates that traditional marine fishers have a reservoir of social capital, which gives them the power of social relations and networks. Such firmly rooted social ties and networks pave the way to build and sustain institutional set-ups. Robert Putnam – one of the leading thinkers of the concept of "social capital" (as noted by Stirrat, 2004) – popularized it by emphasizing social ties, social networks, moral values, and social values in civic and public and policy issues.

> Social capital refers to connections among individuals – social networks and the norms of reciprocity and trustworthiness that arise from them.
> (Putnam, 2000, p. 19)

Putnam (2017), in his recent draft, clarifies that social capital does not merely represent warm friendships and social ties. In contrast, as he argues, social capital includes the extensive benefits of social networks, trust, reciprocity, mutual aid, information, and cooperation. It works through multiple channels like information flows, reciprocity, mutual assistance, collective action, mutual respect, and feelings. Social capital continues to play an important role in the social adaptation to climate change literature (Adger, 2003; Aldrich et al., 2016). Social capital – which encompasses bonding, bridging, trustworthy linkages, reciprocity, and social networks – connects communities and institutions and facilitates collective action (Putnam, 2000), and it is especially relevant in climate adaptation discourses (Adger, 2003; Pelling & High, 2005; Aldrich et al., 2016). In some circumstances, social capital enhances maladaptation and reduces an individual's ability to adapt (Paul et al., 2016). Formal and informal networks develop social capital, relieve poverty, and provide access to resources (Burt, 2000; Richards & Roberts, 1998). Institutions are connected through social networks at all levels and scales. Rotberg (2010) discusses the functions of formal and informal social networks among the inhabitants of a flood-prone village in Bangladesh, as well as how those institutions assisted them in coping with the effects of climate change. In Bangladesh, two recent studies (Islam & Walkerden, 2014,

2015) studied how bonding and bridging networks and social capital spurred disaster resilience and recovery. Despite this, there is scant literature on the role of "social networks" in climate change adaptation in the context of India, particularly regarding "place-based" maritime fishing communities.

The political ecology of climate change adaptation

The impact of poverty and other forms of inequality was recognized early in the climate change debate. However, it has little bearing on the politics, power, and social and economic dominance of certain strata of society (Jayaraman, 2015). Early phases of climate change adaptation research gave limited attention to power, politics, and gender; these dialogues only began to emerge in the 2000s and later. The political ecology discourse provides a lens through which we can examine climate change adaptation in terms of power and politics. Many recent works provide exclusive attention to the ways in which power and politics mediate the individual and collective responses to the changes in environment and climate (Eriksen & Lind, 2009; Manuel-Navarrete, 2010; Pelling, 2011). In discussing climate change adaptation, Eriksen et al. (2015) question the power and politics of implicit adaptation actions: Why do adaptation processes further reinforce the authority of the state, policymakers, and other experts and leave out the vulnerable populations, who are actually the victims of climate change? They conceptualize adaptation "all the way through" and stress that "adaptive is always political and contested." They argue that adaptation processes are not universal, as certain adaptive actions may be advantageous for one group but maladaptive for others. To manage such complications, they argue that adaptation must be explicitly framed as a sociopolitical process and that adaptive processes must be made or viewed as a part of the social dynamics. In other words, adaptation to climate change should not differ from other types of societal and environmental transformations.

"Political ecology" has long been employed to investigate environmental problems in the developing world. Almost 30 years ago, Bryant (1992) emphasized the significance of integrating politics and ecology to understand human–environment relationships and examine environmental degradation. Political ecology comes in various forms, and a sizable body of literature is opposed to it (Rocheleau et al., 1996; Sultana, 2015; Sultana, 2021). Miller and McGregor (2020) examine the global and regional scale for political ecology research and action, providing significant theoretical critiques of general understandings of adaptation, vulnerability, place-based case study research, and abstract policy-focused research. Beyond this, the authors argue for rescaling political ecology to include alternatives and supra-national processes (which are currently missing) to achieve more effective and progressive climate politics through regional research collaboration and networks. Over the past few decades, there has been a growing body of knowledge in the Global South, particularly in India, on the various ways in which power relations and politics influence human–environment relations. Focusing on India, political ecology has been applied to examine conflicts over access to resources, environmental degradation, conservation complexities, developmental issues, and many other environmental problems (a few works: Sanyal, 2006; Truelove, 2011; Doshi, 2017; Cornea et al., 2017; Shrestha et al., 2019; Rai et al., 2019). One recent work focusing on the political ecology of climate change adaptation in the Global South is Marcus Taylor's book *The Political Ecology of Climate Change Adaptation: Livelihoods, Agrarian Change, and the Conflicts of Development*. In that book, Taylor examines climate change adaptation as a development practice and the complex field of knowledge production, particularly focusing on rural regions of India, Pakistan, and Mongolia. Through the cases and theoretical insights, he illustrates climate change adaptation as a fundamentally political process that must be viewed critically and not as self-evident. Taylor clarifies that "climate change adaptation, therefore, is intrinsically a political process despite its pretensions otherwise" and emphasizes the significance of analyzing climate change adaptation through the lens of production and power. Agriculture's commercialization, property rights, capital, state power, migration, technological advancements, etc., are all essential contexts in which to place this phenomenon. Even though this work focuses on agrarian regions, a few of its portions offer rich conceptual insights pertinent to this study, where he stresses the cultural representations and political implications of weather and climate and the importance of taking the conversation about adaptation beyond the dichotomy between climate and society. Overall, Taylor offers an alternative view of climate change, arguing that it is a multi-scalar process involving contesting a wide range of social and political actors and not just a global phenomenon (Taylor, 2014).

Sustainable livelihoods approach

The Brundtland Commission on Environment and Development first proposed the sustainable livelihoods concept in the mid-1980s as a way of improving resource management and production. Initially, however, it was absent from mainstream discussions on poverty eradication, livelihoods, and global development. After the efforts of the international forum Agenda 21 – which highlighted its importance in poverty eradication – the uniqueness of this concept rose to prominence in international development studies (Hoon & Hyden, 2003). Robert Chambers is a pioneer in the field of sustainable livelihoods. The integrated sustainable livelihoods concept, established by Robert Chambers and Gordon Conway, successfully unites the three concepts of capability, equity, and sustainability.

> a livelihood comprises the capabilities, assets (stores, resources, and claims) and activities required for a mean of living: a livelihood is sustainable which can cope with and recover from assets and shocks, maintain and enhance its capabilities and assets, and provide sustainable livelihood opportunities for the next generation; and which contributes net benefits to other livelihoods at the local and global levels and in the short and long-term.
>
> (Chambers & Conway, 1992, p. 6)

Scoones (1998) identifies five types of assets or capitals that are important for achieving sustainable livelihoods. They are, Natural, physical, economic (financial), human, and social capitals are the four types. Some scholars have added a few more "assets" or "capitals" that are important for achieving sustainable livelihoods. They are information and cultural capital (Bebbington, 1999; Odero, 2003; Cochrane, 2006). All "capitals" should be available to the community on an equal basis, as they are crucial in developing resilience and adapting to climate change. The five capitals (natural, physical, human, financial, and social capitals) are unified in the sustainable livelihoods framework, which allows for a holistic understanding of the intricacies of livelihood systems (Farrington et al., 1999; DFID, 2001). Six key principles underpin the concept of sustainable livelihoods: (i) A focus on people, (ii) a comprehensive approach, (iii) adaptive, (iv) capitalizing on your advantages, (v) micro–macro relationships, and (vi) long-term sustainability (Chambers & Conway, 1992; DFID, 2001; Kollmair et al., 2002). According to Hoon and Hyden (2003), it is based on two basic principles: (i). Integrative potential: The concept makes it possible for policies to deal with challenges of economic development, resource management, and poverty reduction all at once. (ii). Its emphasis on the complete complexity of livelihood systems, which must be understood and treated in the context of families, households, and communities, rather than merely jobs. According to Farrington (2001), the sustainable livelihoods method can be investigated on three levels. As a set of principles, they are (i) as a set of principles, (ii) an analytical framework, and (iii) a developmental objective. This approach emphasizes informal institutions in studying livelihoods and minimizing the poor's vulnerability. That is why, when it comes to poverty research, it is critical.

The influence of the sustainable livelihoods approach on climate change adaptation

The sustainable livelihoods approach encapsulates the multiple complexities of livelihoods and discusses how various factors have shaped such complications. It conceptualizes how people operate within the vulnerability context and how they draw the range of institutions (from global to local) to influence the vulnerability of the people (DFID, 2001). This framework identifies and depicts the relationship among the "assets/capitals" of the stakeholders and enunciates how it can enhance/constrain their livelihoods systems in responding to the vulnerability context.

In the words of Robert Chambers, vulnerability thus has two sides:

> An external side of risks, shocks, and stress to which an individual or household is subject; and an internal side which is defenselessness, meaning a lack of means to cope without damaging loss.
>
> (Chambers, 1989, p. 1)

As a result, a variety of circumstances, such as climate change and risks, political instability, famines, natural resource decrease, governance failure, and so on, locate the households'/community's vulnerability context. Over the last two decades, studies on rural development, rural livelihoods, fisheries livelihoods, environmental management, and climate change adaptation have increasingly embraced the sustainable livelihoods framework. This method is advantageous in understanding the vulnerability and adaptation of socioecological systems to climate change in various ways because it examines both the components and the factors that influence livelihoods (Scoones, 1998; Reed et al., 2013). In examining the intricacies and vulnerabilities of small-scale fishers, a holistic and people-centered approach to sustainable livelihood principles is beneficial (Allison & Ellis, 2001; Allison & Horemans, 2006; Islam, 2013; Islam et al., 2014a, 2014b).

Sustainable livelihoods approach – some criticisms

Despite the potential of the sustainable livelihoods approach, it is not free from criticism. A few of them are the following: In this approach, people are invisible and "capital" assets are unclear. The ambiguous operationalization of this approach for assessing capital assets is criticized by development practitioners. Furthermore, this approach has been critiqued for being overly focused on "micro" levels and disregarding the role of markets, major actors, wealthy players, policy environment, higher levels of governance, and so on. The main critique is that this approach is too "micro" and ignores power interactions at greater levels (Moser & Norton, 2001; Small, 2007). Farrington (2001) points out that this approach has some problems when it comes to putting it into action, especially because state-level institutions are so complicated. Consequently, it necessitates greater freedom and adaptability in implementation, which is usually difficult under a conventional institutional structure. Moreover, this method may overlook the significance of intra-household relationships, power, and politics in some circumstances.

Institutions

In his seminal work, North (1991) defines institutions as

> Institutions are humanly devised constraints that structure political, economic, and social interaction. They consist of both informal constraints and formal rules.
>
> (North, 1991, p. 97)

Scott's (1995) definition of institutions captures the enabling qualities of the institutions. He defines

> Institutions consist of cognitive, normative, and regulative structures and activities that provide stability and meaning to social behaviour.
>
> (Scott, 1995, p. 33)

There have been extensive discussions on "institutions" in social sciences literature over the last three decades (Coleman, 1987; North, 1991; Ostrom, 1995; Pahl-Wostl, 2009). Nonetheless, scholars remark there is no "standard" definition to define "institutions" (Jentoft, 2004; Hodgson, 2006). Jentoft (2004) identifies the commonly accepted quality of an institution amongst all, which is durability.

Institutions are divided into three categories: public (bureaucratic administration, local governments), civic (community-based groups, membership organizations, cooperatives), and private (service organizations and private businesses). Local institutions might be either traditional or modern. They are, nonetheless, deeply rooted in the community. In most cases, they are in specific regions, and the people who are involved in the local institutions design the rules and regulations of the local institutions. Some typical local-level institutions are the following: customary village councils, community-based organizations, community mobilization councils, disaster management committees, conflict resolution committees, pastoral committees, livestock and agricultural production cooperatives, religious associations, recreational clubs, youth clubs, security management councils, village asset production councils, religious associations, lineage organizations, traditional authorities, women's groups, local government authority, etc. (Agrawal et al., 2009; Agrawal & Perrin, 2009). Helmke and Levitsky (2004) argue that the rules of formal institutions are explicitly codified, but the rules of informal organizations are unwritten and wholly operated through non-formal channels. They also underline that "informal institutions" are not weak institutions and should be distinguished from "informal organizations" (Helmke & Levitsky, 2004, for more discussions). Pahl-Wostl (2009) expresses a similar point of view, in which institutions are not synonymous with organizations. The formal or informal nature of institutions is determined by the nature of codification, development processes, codification, communication, and enforcement. In general, formal institutions are formally/legally codified and enforced through legal procedures. Informal institutions are not normally codified, and they share/reflect societal social and cultural norms.

Local social institutions in coping and adaptation to climate change

Discussions on local institutions are highly pertinent in responding to climate change impacts because of the following: (i) it regulates and governs human–environment interactions; (ii) it structures the power and rights of the community; it mediates the communities and external resources at multiple levels of governance; (iii) it co-evolves with the changes in the environment; (iv) it mediates the social responses and external agencies into localized contexts, thereby promoting the community's adaptation efforts; (v) it structures the distributions of the risks of climate change; (vi) in most cases, local communities and their surrounding institutions have already experienced the differences in their local environment and its consequences; (vii) the key leaders of the local institutions influence the climate adaptation efforts at the local levels; (viii) local institutions are firmly rooted in traditional communities (for example, traditional marine fishers), and they are already well-positioned to deal with environmental threats and projects; and (ix) local institutions are the fundamental mediating

Conceptual underpinnings and introducing the framework 31

mechanisms to convert external interventions into adaptation to climate change (Bakker et al., 1999; Agrawal, 2008; Agrawal & Perrin, 2009; Berman et al., 2012; Marschke et al., 2014). In terms of formal and informal norms, three types of local institutions are relevant to climate adaptation: civic, public, and private (Donnelly-Roark et al., 2001; Agrawal et al., 2008; Agrawal & Perrin, 2009). Institutions impact community decision-making and can support collaborative climate change action. Globally, a growing body of research emphasizes the importance of local institutions in climate change adaptation (Termeer et al., 2012). Institutional settings affect the household economy across and within scales. It allows collective effort to face climate change issues (Gentle et al., 2013).

Similarly, Petzold and Ratter (2015) stress the necessity for intensive political and cooperative efforts to adapt to climate change impacts, especially sea-level rise. Understanding the social characteristics of the community and how they are intertwined with local decision-making processes is required for evaluating the efficiency of adaptation strategies in addressing the challenges of sea-level rise. Local institutions help us comprehend the power and governance arrangements at various community levels. For effective governance, formal and informal institutions should be compatible and complementary (Pahl-Wostl, 2009). Mubaya and Mafongoya (2017) categorization enables us to understand the two distinct realms of literature on climate change adaptation and institutions. According to them (Mubaya & Mafongoya, 2017), Adger's (2006) work on the institutional framework is the classic representation of the first domain of institutions literature; it claims that climate change vulnerability and adaptation are primarily shaped by sociopolitical factors and institutions, not merely biophysical elements. In comparison, Agrawal and Perrin's (2009) study on local institutions in climate change adaptation is an excellent illustration of the second domain of literature since it focuses on the roles and actions of local institutions in climate change adaptation. This study adopts Agrawal et al. institution's typology (2009).

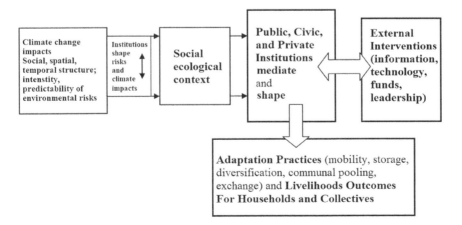

Figure 2.2 Adaptation, Institutions, and Livelihoods framework.
Source: Agrawal, 2008.

32 *Conceptual underpinnings and introducing the framework*

In the study of the local institutions in rural Ethiopia and Mali, Crane (2013) finds that there is a high likelihood of overlap and competition between the newly constructed local institutions and the customary (traditional) institutions in performing the tasks and in capturing the natural resources; this may impede the adaptive ability of the society. Moreover, the power layers of traditional communities are not readily obvious but are inextricably interwoven with the community's cultural norms and practices; these power layers may compete with external organizations and weaken their credibility.

Local institutions in the face of climate change

At the local level, multi-level social networks, formal and informal local institutions, and social learning can help communities cope and adapt (Adger et al., 2005a, 2005b; Osbahr et al., 2010). Despite the importance of understanding the scope and limits of local institutions, there have been huge knowledge gaps in understanding the roles of formal and informal institutions in shaping, building, and sustaining the adaptive capacity of the local communities to climate change (Keys et al., 2014), and the role of social networks in promoting resilience has been under-researched in the Global South (Rockenbauch & Sakdapolrak, 2017). Focusing on local institutions provides us with a lens through which we can better understand local-level processes, diversity of local actors, institutional relationships, community self-autonomy, and mechanisms of local-central articulations (Agrawal et al., 2009), all of which are critical to understanding climate adaptation actions of self-governing communities. However, it is important to recognize that local institutions provide the best results only in certain settings, and that it is not reasonable to expect major results in societies with deeply ingrained informal norms and regulations and where state institutions exert greater control over local-level management. Formal and informal institutions are key to facilitating collecting efforts and developing social networks to address environmental risks. Similar social identities allow communities to adapt to climate change through social relationships and networks (Agrawal & Perrin, 2009; Pelling, 2011).

Conceptual framework

This book adopts both the sustainable livelihoods framework (Chambers & Conway, 1992; DFID, 2001) and the AIL framework (Agrawal & Perrin, 2008) to address the investigative questions. The livelihoods framework (Chambers & Conway, 1992; DFID, 2001) illustrates the interlinking of different assets/capitals among one another. Allison and Horemans (2006) stated that the sustainable livelihoods framework is applied to social groups/ households – where they reside together in the same place, and collective decision processes coordinate their activities, incomes, and resource sharing. In recent decades, several scholars have applied the sustainable livelihoods approach to investigate the management, assets, survival strategies and impacts, vulnerability, and adaptation actions of fishing communities to climate change (Allison & Horemans, 2006; Divakarannair, 2007; Badjeck et al., 2010; Islam, 2013; Islam et al., 2014a, 2014b; Deb & Haque, 2016). However, such investigation of small-scale fisheries using the sustainable livelihoods approach is very limited in the case of coastal South India, particularly on the Coromandel coast of Tamil Nadu. Only scanty research knowledge is available to understand the climate change adaptation actions of small-scale marine fishers of South India. Within that, the research with theoretical back-up in investigating marine fisheries is very scanty. Across the world, community assets/capitals of fishing communities were categorized by adopting the early literature (for example, Ellis, 2000; Allison & Ellis, 2001) on sustainable livelihoods (Divakarannair, 2007; Badjeck, 2008; Islam, 2013; Momtaz & Shameem, 2015). Divakarannair (2007) extends this discussion and adds one more type of assets/capital – political assets, which is particularly relevant to this research as it intends to focus on the local institutions and capitals/assets of the marine fishers (see Table 2.1).

The AIL framework by Agrawal and Perrin (2008) discusses the roles of institutions in adaptation to climate change. In recent years, several authors used the AIL framework to analyze the roles of institutions in adaptation to climate change. For example, Berman (2014) adopts a polycentric approach to studying the roles of informal institutions in climate adaptation by embracing the AIL framework of Agrawal and Perrin (2008). Wang et al. (2016) adopt the AIL framework to discuss climate adaptation, local institutions, and sustainable livelihoods of herder communities in northern Tibet. By

Table 2.1 Capitals/assets of marine fishers

---
i. Natural capital – Marine fisheries resources, coastal spaces, water, fisheries resources in the river.
ii. Physical capital – It denotes the infrastructure and assets of the fishing communities, for example, Fishing vessels (Catamaran, FRP boats, other big fishing vessels, fishing gears and nets, purse nets (surukku valai), navigation instruments, houses, sanitation, technology, public infrastructure, markets, transport, and communications.
iii. Human capital – It includes fishing skills, technical knowledge, indigenous knowledge, health, education, labor power, ability to fish, strength to diversify livelihoods, and nutrition.
iv. Financial capital – It includes cash, savings, jewels, cash remittances from their family members, other external financial sources, and immovable property assets.
v. Social capital – It refers to the social networks and relationships of the fishers, which makes them diversify their livelihoods strategies, and supports them during crises. It includes trust, mutual reciprocity, kinship relations, friendships, and informal safety nets.
vi. Political capital – Political capital mainly refers to the active involvement of political parties by marine fishers. It includes the fishers' membership and their active participation with the parties.

Source: Adapted from Ellis, 2000; Allison & Ellis, 2001; Divakarannair, 2007; Badjeck, 2008; Islam, 2013.

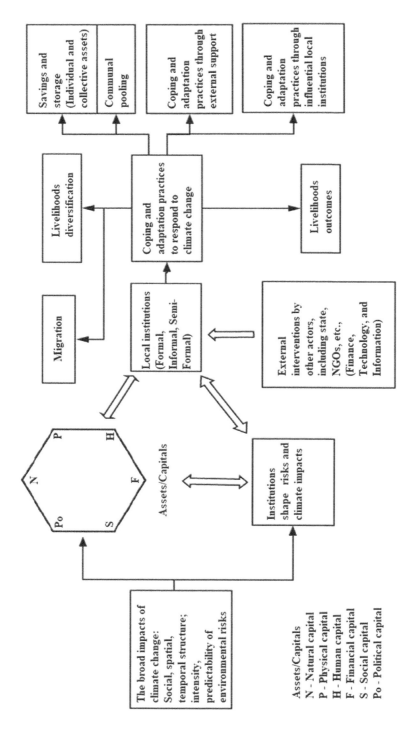

Figure 2.3 "Assets/Capitals based Adaptation, Institutions, and Livelihoods (AIL) framework."
Source: Adapted from DFID, 2001; Agrawal and Perrin, 2008.

taking this forward, the present study attempts to interlink the sustainable livelihoods framework with the AIL framework (Agrawal & Perrin, 2008) and propose the "Assets/Capitals based Adaptation, Institutions, and Livelihoods (AIL) framework" as it intends to investigate the ways in which the local institutions mediate the responses of marine fishers in adapting to climate change. This framework attempts to explore the synergies between them, develop an integrated framework, and discuss how the formal, informal, and semi-formal local institutions of self-governing small-scale fishing communities can interact with one another in response to climate change. It (Figure 2.3) depicts how the institutions and capitals/assets mediate the impacts of climate change. The nature of assets is intrinsically linked. Hence, the assets hexagon has been developed to understand and analyze the livelihoods vulnerability of the marine fishers in the face of climate change. Here, I remind you that the sixth asset/capital (political capital) that I included in the asset hexagon mainly refers to the political parties.

This framework recognizes both the assets/capitals and local institutions (formal and informal) shape the livelihoods of the marine fishers by mediating the vulnerability of climate risks and change. Focusing on the local institutions in the framework addresses the weakness of SLA – which is mostly centered on its inability to analyze the power and power relations of the stakeholders. By mediating six assets/capitals and the local institutions, the marine fishers respond to the vulnerability context to create various coping and adaptation practices, including livelihoods outcomes, livelihoods diversification, migration, and adaptation practices through external support and influential local institutions.

3 Unraveling climate change and disasters

Navigating to the field

Introducing climate change and its impacts along the Coromandel coast of South India

> No common theoretical or methodological framework for understanding fishermen and fisheries has been developed in social science, but this does not mean that the area has not been studied. Several social science disciplines have contributed to the research done regarding fishermen from different perspectives.
>
> (Christensen, 2007, p. 6)

Patnaik and Narayanan's (2005) research on the vulnerability of India's eastern coastal districts to climate change indicates that climate change will have a variety of negative effects on poverty-reduction initiatives, and physical and social infrastructures of coastal communities. There has been a growing body of research evidence over the past decade recognizing the vulnerability of India's coastal regions to climate change and coastal hazards (Saxena et al., 2013; Chakraborty & Joshi, 2014; Ramachandran et al., 2016; Geetha et al., 2017; Rao et al., 2020; Rehman et al., 2021). The recent review on climate risks and impacts shows that coastal megacities in India are potentially vulnerable to sea-level rise, temperature rise, increasingly intense tropical cyclones, and riverine flooding. Two Indian coastal cities, Kolkata and Mumbai, are among the most vulnerable to climate change impacts in South Asia (Vinke et al., 2017). The climate of India lies within the tropical monsoon zone, which indicates the high spatial variation in rainfall and its frequencies. It also indicates the impacts of its location lie within the tropical belt (Guhathakurta et al., 2011; Jaswal et al., 2015). Brenkert and Malone (2005) analyze the vulnerability to climate change by using Vulnerability-Resilience Indicator Prototype (VRIP) and find that six coastal states (Goa, West Bengal, Kerala, Tamil Nadu, Odisha, and Gujarat) are more vulnerable to climate change than other states of the country. All these states are highly densely populated. Further, two coastal states (Odisha and Tamil Nadu) show high sensitivity to storm surges. Guhathakurta et al. (2011) analyze rainfall data for 1901–2005 across India and find that

DOI: 10.4324/9781003187691-3

rainfalls significantly vary across the regions and highlight the significant increase in flood risks across India in the past century.

Recent scientific research suggests that the Indian monsoon rainfalls and temperatures have become increasingly anomalous over the years and decades. It highlights the increasing trend of changes in weather and climate systems across India (Jaswal et al., 2015; Ghosh et al., 2016; Guhathakurta & Revadekar, 2017; Mukherjee et al., 2018). According to Jaswal et al. (2015), on examining India's hot summer days from 1969 to 2013, hot summer days in South India increased by around 18%, and hot summer days in North India climbed by nearly 18% between 1999 and 2013. Such increasing patterns of warm days throughout the years have had profound impacts on human and natural systems. It demonstrates that the number of hot days varies greatly across India's eastern and southern areas, particularly over Gangetic West Bengal, Odisha, and Tamil Nadu. Temperature and heatwave patterns are predicted to rise across India. Heatwaves and high temperatures will have a greater influence in the future, particularly in numerous Central, North, and South Indian regions. Such escalating tendencies would certainly lead to an increase in humidity and sea surface temperature, which will have several negative repercussions on human health (van Oldenborgh et al., 2018).

Small-scale marine fishers in India have encountered a variety of challenges in securing their livelihoods during the last three decades (Korakandy, 2008). Industrial fishing, aided by state-sponsored neoliberal policies, has made them more vulnerable (Nandakumar & Muralikrishna, 1998; Mathew, 2009; Panigrahi & Mohanty, 2012; Bhagirathan et al., 2010). Bhathal (2014) identifies that the fish capture of Indian fisheries has hit its natural limits and is no longer sustainable. The massive growth in the number of fishing vessels over the past few decades has exerted enormous strain on India's traditional fishing grounds. Several recent research studies show the deterioration of marine fishing grounds in India as a result of an increase in mechanized fishing vessels. Increasing temperatures and erratic changes in weather and climatic patterns further aggravate the decline of fish productivity in India (Vivekanandan, 2010; NCAER, 2010; Salagrama, 2012; Kizhakudan et al., 2014; Zacharia et al., 2016).

Coastal Tamil Nadu

Tamil Nadu, the southernmost state of the Indian peninsula is in the Northern hemisphere in the torrid zone between 8°5′ to 13°35′ N and 76°15′ to 80°20′ E and accounts for around 4% of the total area of the country. Tamil Nadu receives rainfall from Northeast and Southwest monsoons, and a significant share comes from the Northeast monsoon. Across Tamil Nadu, the temperature remains relatively humid throughout the year (Bal et al., 2016). The average rainfall in Tamil Nadu is about 950 mm/year and the average number of rainy days is about 50 a year. However, Tamil Nadu is considered a water-deficient state (SoE, 2005). Recent research examines

Table 3.1 Vulnerability of coastal Tamil Nadu to climate change and extremes

Authors	Findings and discussions
Khan et al. (2012a)	Using a participative approach (stakeholder analysis and rapid rural appraisal), this study analyzes the vulnerability of mangrove-dependent communities in Pichavaram forests to sea-level rise in southeastern coastal Tamil Nadu. They develop a methodological framework to bring awareness among coastal people (especially fishers and farmers).
Sheik Mujabar and Chandrasekar (2013)	Rapid changes in geology and geomorphology, as well as sea-level change, tropical cyclones, and storm surges, threaten India's southeastern coastal Tamil Nadu (which includes Nagapattinam district).
Bal et al. (2016)	Using the Met Office Hadley Centre regional climate model, this study examines regional climate change projections in Tamil Nadu. It concludes that the state's average temperature is increasing over the baseline period (1970–2000) and that yearly rainfall projections for the same period are decreasing.
Chakraborty and Joshi (2016)	By rating composite indicators, this study examines the vulnerability of Indian districts to natural (earthquakes) and climate-induced disasters (cyclones, floods, droughts, and sea-level rise). It reveals that coastal Tamil Nadu is especially vulnerable to sea-level rise and cyclones; the Nagapattinam district is one of them.

rainfall variability and changes in rainfall patterns across the state from 1989 to 2018 and finds that the Nagapattinam district shows decreasing trends of rainfall patterns over the years (Guhathakurta et al., 2020). Based on rainfall patterns, irrigation, agriculture, soil features, and other social and ecological features, Tamil Nadu is divided into seven agro-climatic zones: (i) Cauvery Delta, (ii) Northeastern, (iii) Western, (iv) Northwestern, (v) High Altitude, (vi) Southern, and (vii) High Rainfall. The rainfall patterns vary across the regions of the state as well as seasons (Bal et al., 2016). The Nagapattinam district, which is the subject of this research, is in Tamil Nadu's Cauvery Delta. Nagapattinam has a semi-arid climate. This district receives around 70% of its yearly precipitation from the Northeast monsoon and about 15% from the Southeast monsoon. The yearly mean temperature in this district is 28.79°C, while the mean maximum and lowest temperatures are 32.66°C and 24.9°C, respectively (Kumaraperumal et al., 2007).

Nagapattinam (Or Nagai) district: deep-rooted cultural history, abundant resources, and climate extremes

Tamil Nadu's coastal waters are famous for their abundant fishing resources. The Nagapattinam coast is home to 95 different fish species from 42 families and 59 genera (Ramu et al., 2015). The coastal waters of Nagapattinam are renowned for their abundance of Mackerel, Seerfish,

Oil Sardines, Flying Fish, and Anchovies. An estimated 90,637.3 tons of marine fish were produced in the Nagapattinam district during the 2018–2019, with 32,249.69 tons coming from the non-motorized sector and 459.8 tons from the non-mechanized sector, respectively (Government of Tamil Nadu, 2019). Due to extreme storm surges and storm surges riding over tides and cyclones, Tamil Nadu experiences coastal flooding regularly. From 1900 to 2004, Tamil Nadu's coasts were battered by cyclonic storms around 30 times, each of which had devastating consequences. The affected districts are Chennai, Cuddalore, Nagapattinam, Thanjavur, Ramanathapuram, and Kanyakumari (Sundar & Sundaravadivelu, 2005). Due to their geographical features, these districts are vulnerable to tropical cyclones, storm surges, and the associated climate risks and change. Since the disastrous impacts of the 2004 Indian Ocean Tsunami disaster, the Tamil Nadu state government has been keen to protect the lives and livelihoods of coastal communities. It has implemented many welfare schemes to protect the lives and livelihoods of coastal fishing communities. Some of them are as follows: (i) Emergency Tsunami Reconstruction Project (ETRP) – World Bank's financial assistance (Global Facility for Disaster Reduction and Recovery, GRDRR, World Bank, 2006); (ii) Tsunami Emergency Assistance (Sector) Project (TEAP) – With assistance from the Asian Development Bank, its main objectives are to restore the coastal livelihoods and to increase the infrastructure and capacity building of coastal districts (Government of Tamil Nadu, 2007; ADB, 2012); (iii) vulnerability reduction of coastal communities (VRCC) is a project backed by the World Bank. Its main goal is to help people in areas hit by the tsunami get back on their feet and get back to work (World Bank, 2007); (iv) Post Tsunami Sustainable Livelihood Programme: This is an IFAD-assisted project, and its main objective is to build self-reliant coastal communities (IFAD, 2005).

A growing body of research evidence shows that coastal Tamil Nadu, particularly, the Nagapattinam district,[1] is extremely vulnerable to coastal disasters and climatic extremes (Bhalla et al., 2008; Byravan et al., 2010; Khan et al., 2012b; Arivudai Nambi & Sekhar Bahinipati, 2013; CSTEP, 2014; Ramachandran et al., 2017). The low-lying area encompasses the majority of the eastern coast north of the Nagapattinam district. Five districts in Tamil Nadu, including Nagapattinam, are extremely vulnerable to sea-level rise. The majority of the Nagapattinam district lies either below or above sea level. For this reason, it is more vulnerable to flooding from the sea (Mascarenhas, 2004; Byravan et al., 2010; Arivudai Nambi & Sekhar Bahinipati, 2013). The sensitivity and exposure of the Nagapattinam district have made it more vulnerable to climate change impacts (CSTEP, 2014). These coastal areas are also prone to flooding and extreme temperatures during the summer (Bal et al., 2016). The extreme climate events and the resulting damage to the Nagapattinam district are summarized in Table 3.2.

Table 3.2 Nagapattinam district: a saga of climate extreme events

Month and year	Climate extreme events	Loss of lives and assets
1 Dec 1984	Floods due to heavy rain	Crops were damaged on a large scale and affected normal life due to heavy floods
15 Nov 1991	Heavy rainfall	Crops damaged
04 Dec 1993	A severe cyclone with a speed of around 188 km/h	Thousands of people in Tamil Nadu and Karaikal regions lost their livelihoods
26 Dec 2004	Tsunami disaster	6065 fatalities, 791 missing, 1922 injured, big damages to infrastructure and groundwater resources
27 Nov 2008	Nisha cyclone, speed 80 km/h	20 Life Loss, 4,58,949 houses were damaged
Nov 2010 & Dec 2010	Heavy rains	10 Life loss, 1492 cattle loss, big damages to crops, houses, and groundwater resources
31 Dec 2011	Thane cyclone	Hut damages partly 1468, Fully 24, seawater intrusion into the fishing villages
9 Nov 2015 & Oct–Dec 2015	Massive rains	Heavy damages to the infrastructure and fishing assets, acute drinking water shortage for one month, and damages to groundwater
16 Nov 2018	Gaja cyclone	63 fatalities, 3.7 lakh people homeless across the state. It shattered the livelihoods of millions and brings back the memories of the 2004 Indian Ocean Tsunami disaster. Acute water shortage in the district along with severe damages to groundwater resources (NDTV, 2018a; NDTV 2018b)

(Source: Nagapattinam District Disaster Management Report, 2012).

The consequences of the 2004 Indian Ocean Tsunami on the Nagapattinam district

The Nagapattinam district on the Indian mainland was the worst hit by the 2004 Indian Ocean Tsunami disaster. The damage from the tsunami was made worse by the district's social and geographical vulnerabilities. Infrastructural damage, water resource degradation, agricultural losses, fishing losses, and damage to coastal resources were all long-term implications of this catastrophe (Sheth et al., 2006; Kume et al., 2009). The flat topography of the Nagapattinam district is the most vulnerable factor to seawater inundation, according to Ramana Murthy et al. (2012) research on the effects of the Tsunami disaster on the Nagapattinam coast using post-tsunami field measurements. There was a wide variation in the extent of the damage along these coasts due to factors such as land topography, coastal wetland areas, coastal geomorphology, and tsunami wave propagation.

This disaster, however, ushered in various economic opportunities and social advancements for the district's coastal fishers. As part of rehabilitation measures, international and national NGOs built (mostly) free houses for fishers under the humanitarian aid agenda (Chandrasekhar, 2010; Gandhi, 2011; Swamy, 2011). The inflow of funds from different donors to this district for recovery and rehabilitation activities was so huge, as Esther Fihl remarks, "the second tsunami" (2014). The issues spanning housing-reconstruction, relocation of the fishing community, and free housing between the state and the fishers of Nagapattinam district were extremely complex (Swamy, 2009; Swamy, 2014; Lakshmi et al., 2014). The government, foreign NGOs, and donor organizations provided free housing and various rehabilitation services to tsunami victims. The state regulated it several times. Nonetheless, as various researchers have noted, the state's roles were frequently insufficient (Chandrasekhar, 2010; Swamy, 2011; Lakshmi et al., 2014). As Babu (2011) notes, private players played a big role in the post-tsunami activity. Since then, Tamil Nadu's coast has emerged as a desirable location for industrialization, bolstered by large-scale investments (Babu, 2011; Swamy, 2011). It has had a significant impact on the rising levels of pollution on the beaches and among marine fishers.

Research method, field, and techniques

Fisheries, as described by Mahon et al. (2008), are "complex adaptive human-in-nature systems." Given the complexities of small-scale fisheries systems, this study employs a qualitative data collection technique through a case study.

> Case study is an empirical inquiry that examine a contemporary phenomenon in its real-life context, especially when the boundaries between phenomenon and context are not clearly evident.
>
> (Yin, 2003, p. 13)

The case study is a flexible research approach that enables researchers to investigate complex real-world phenomena while preserving underlying holistic and essential qualities (Schell, 1992). The research goals of this study tend to investigate in a real-world setting, where respondents' subjective experiences, perceptions, and interpretations, as well as the researcher's, play a critical role. Ford et al. (2010) examine the history of using case studies in climate change research and argue in favor of case study research approaches. He advocates the significant need for doing case study research to understand climate change impacts and societal responses to climate change. To begin the study process, it is necessary to get to know the fishermen and the daily obstacles in their lives and livelihoods (Christensen, 2007). As a result, I conducted this research using a qualitative case study method (Yin, 2003). The constructivist grounded theory technique (Glaser & Strauss, 1967; Corbin & Strauss, 1998) also focuses on how small-scale marine fishermen and their local institutions comprehend, interpret, and adapt to climate change in this study. Grounded theory is mostly inductive and seeks to create the theory through data collecting (Morse, 2001).

> The researcher does not begin a project with a preconceived theory in mind... Rather, the researcher begins with an area of study and allows the theory to emerge from the data.
>
> (Corbin & Strauss, 1998, p. 12)

According to Corbin and Strauss (1998, p. 12), research should not begin with any preconceived theory or mindset. Whereas, as stated in previous chapters, this study focuses on sustainable livelihoods theory (DFID, 2001) and the AIL framework (Agrawal, 2008). As a result, while the constructivist grounded theory method inspired this study, it was not employed in its purest form. This study was mostly conducted using qualitative methodologies.

Navigating to the field

Midway through 2014, I conducted pilot surveys in nine fishing villages in the Nagapattinam district. As part of the pilot survey, I examined scientific research papers and field reports from local non-governmental organizations to understand the unique circumstances, vulnerabilities, and potentials of fishers in this region. I directly visited the government offices of this district and gathered reports on the impacts of the 2004 Tsunami disaster, the 2011 Thane cyclone, and the 2015 heavy rains on the fishing villages in this district. During this period, I also met with a few local experts who are knowledgeable about climate change and its risks and effects on this district. In the pilot survey, I did not limit myself to small-scale fishing villages. As a result, in addition to small-scale fishers, I interacted with marine fishers and other stakeholders from fishing villages of this district that are primarily engaged in mechanized fishing. Overall, the fishing villages I studied in Nagapattinam district were Aarukkattuthurai, Kodiyakkarai, Kodiyakkadu, Pushpavanam, Vellappallam, Vanavanmadevi and Akkaraipettai, Kodiyampalayam and Koozhayar. Exclusive of Kodiyampalayam and Koozhayar, all the other fishing villages are predominately engaged in mechanized fishing.[2]

In the pilot survey, fishermen from Kodiyampalayam and Koozhayar expressed their concerns about mechanized fishing and its detrimental consequences on fisheries. Along this coastal region, fishers have substantial marital and kinship relationships, leading to self-organization and integration. It enables them to resolve their own disputes without the intervention of other entities. The fishers of this stretch are bound together by their extensive kinship relationships and robust internal governance systems. As a result, the use of banned nets has not emerged as an inevitable problem among fishing communities, even though it frequently causes conflicts among them. The caste diversity is visible in some "big"[3] coastal villages (for example, Kodiyakkarai, Kodiyakkadu, Vanavanmadevi) in "big" fishing villages. Apart from Pattinavars,[4] *Mutharaiyar*,[5] *Adivasi* (Scheduled Tribe), *Pandaram* (Other Backward C), *Muslims*, *Adidravidar* (Scheduled Caste), and other castes, most of which belong to "Most Backward Castes," have settled in this region for years. The government of Tamil Nadu lists Pattinavars (both Periya Pattinavars and Chinna Pattinavars[6]) as Meenavars (fishers) and puts them in the "Most Backward Caste" group. Pattinavars are so eligible for state reservation quotas in terms of education and job prospects (Government of Tamil Nadu, 2022b). The diversity of castes and occupations among the fishers also act as the factors that loosen the tight-bonding and self-governance features of these villages, unlike the homogenous fishing villages. The pilot survey also reveals that the livelihood challenges of these fishing villages are broad and are not limited to climate change risks. Many fishers in these villages (particularly Kodiyakkarai) are, for example, victims of Sri Lankan Navy personnel that put their lives and livelihoods at risk. This study does not cover discussing geopolitical problems or how mechanized fishing could hurt marine resources as those issues lie beyond the scope of this book. In the pilot survey of this study, I concentrated on the social bonding and self-organization ability of fishing villages in response to climate change –specifically, my attention was on fishing villages mostly involved in small-scale fishing.

In the case of small-scale fishing villages, I found that social networks and local institutions had a significant impact on the fishers. The following are some reasons. (i) Low population: As a result of the low population in these villages, the fishing households have developed strong social networks. They frequently meet and converse in public places. (ii) Single-caste: The homogeneous feature of single-caste fishing hamlets links the fishermen more tightly than in diverse fishing villages. Pattinavars and non-Pattinavars are less likely to engage in conflict since the internal governance systems of the small fishing villages are highly self-organized. (iii) Less occupational diversity: Less occupational diversity among the fishers provides no alternative livelihood options other than a marine fishery. "If you are born as a fisher, you must do fishing" is an unwritten rule in fishing villages that are primarily involved in small-scale fishing. No viable alternative livelihoods options integrate the fishers to confront their problems collectively. (iv) The apathy of state government and external agencies toward these villages also makes them stand united.

I attempted to obtain village maps from the local panchayat offices but was often unsuccessful. They informed me that they had misplaced the maps, and it is assumed that the local actors were unaware of the significance of the village maps. Employees of local governments furnished me with maps of the hunger project in some situations (2008). I also examined CMFRI Census data, Nagapattinam District Disaster Management reports, and local reports from SIFFS, IFAD, SNEHA (NGO), ICSF, and MSSRF to better understand how marine fishers in Nagapattinam district reacted to the 2004 Indian Ocean Tsunami disaster, climate change, and overfishing. In June and July of 2015, I did fieldwork in three more fishing villages in the Nagapattinam district that are vulnerable to climate change and risks, in addition to the aforementioned fishing villages, in order to understand the formal and informal governance systems of local institutions by interacting with local NGO employees and two government officials. The following points were discussed: (i) Over-exploitation of marine resources and its impacts on small-scale fishers; (ii) the long-term effects of the 2004 Indian Ocean Tsunami disaster; (iii) the recent weather and climate change events and their implications on fishers; (iv) the sources of early warning systems; (v) the capacity of the fishing villages to adapt to climate

44 *Unraveling climate change and disasters*

change; (vi) roles and activities of the local institutions and their influence over the fishing villages; (vii) potentials and limitations of local institutions, with an emphasis on the complexities and contests amongst fishermen in the sharing of marine resources and maritime territories.

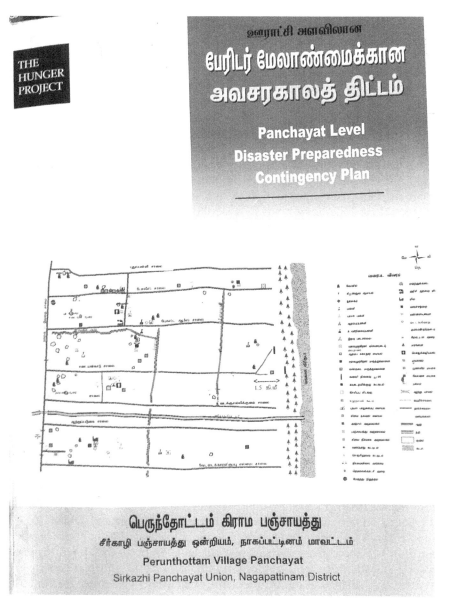

Figure 3.1 The front cover of The Hunger Project report (Perunthottam gram panchayat where Chavadikuppam is located).

Source: The Hunger Project.[7]

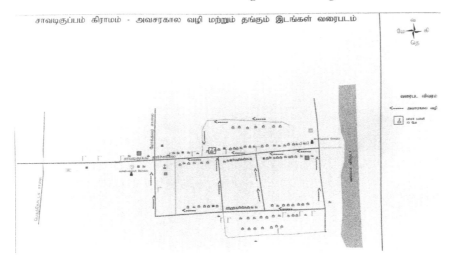

Figure 3.2 Evacuation route planning of Chavadikuppam in disaster situations (a sample).
Source: The Hunger Project.

Selection of the study villages

The pilot survey and second phase of fieldwork found that the small-scale fishing villages in this district are highly vulnerable to climate change for the following reasons. (i) There are no feasible alternatives for the fishers in these communities, which puts them at greater risk during and after extreme weather conditions. (ii) Many of these fishing villages along this stretch of coast are not protected by seawalls. However, they are bordered by fishing villages where mechanized fishing predominates; interestingly, these villages are protected by seawalls. The fishers observe that the restrained sea waves of these mechanized fishing villages impacted the coasts of their villages with more power. So, their villages are prone to erosion and seawater intrusion. (iii) Regardless of fishers' social, political, and institutional backgrounds, the 2004 Indian Ocean Tsunami calamity left a severe scar on the fishers' psyche. Long-term consequences of the tsunami continue to combine with the rising effects of climate change, putting small-scale fishers at risk. In light of these considerations, small fishing villages remained a focus of this research because they offer a wide range of possibilities for examining the research aims.

Defining "small" and "big" fishing villages

Small-scale fisheries are difficult to define because their definitions vary by region. Several academics and institutions define what constitutes small- and large-scale fisheries (Panayotou 1982; World Bank, 1991; Kurien, 1998; Berkes et al., 2001). As FAO defines artisanal fisheries

46 *Unraveling climate change and disasters*

Figure 3.3 Tsunami memorial tombs in the study villages.
Source: Picture credit: Devendraraj Madhanagopal.

> Traditional fisheries involving fishing households (as opposed to commercial companies), using relatively small amounts of capital and energy, relatively small fishing vessels (if any), making short fishing trips, close to shore, mainly for local consumption.
>
> (FAO, 1999)

This study takes into account Kurien's (1998) definition of small-scale fishing and Johnson's conceptual explanations of small-scale fishing systems (Johnson 2018). Small-scale fisheries have the following characteristics: (i) Use small fishing vessels and simple crafts with low capital intensity; (ii) skill-intensive fishing activities; (iii) relying on local traditional knowledge to predict weather and climate patterns for fishing; (iv) nearshore fishing is limited to a single day; (v) Influence of informal leaders and informal institutions in fisheries management; (vi) socially, economically, and politically disadvantaged sections (Kurien, 1998). Hence, by analyzing the geographical and social vulnerabilities of the fishing villages and by considering Kurien's (1998) definition, I sorted out seven fishing villages of Sirkazhi Taluk of the Nagapattinam district of the Coromandel coast of Tamil Nadu to conduct this study. Nevertheless, considering the local context, I applied the term "small-scale fishing" flexibly in the study sites to include two fishing villages where relatively fewer households are engaged in "surukku valai" (purse net)[8] fishing. Bycatch in purse-seine fisheries is poorly understood in the context of India and needs to be studied more thoroughly. The purse seine method is one of the most advanced commercial fishing techniques for catching shoaling fish. Compared to small-scale fishermen (including those who use motorized vessels), purse-seine fishermen are more likely to utilize advanced fishing equipment such as Global Positioning System (GPS), communication equipment, and (VHF marine transceiver and mobile phones) to navigate and facilitate their fishing operations at sea. On the Coromandel coast of South India, surukku valai fishing is predominantly used to capture shoaling fishes such as sardines, mackerels, etc. (Mariappan et al., 2017; Colwell et al., 2019). Through a micro-study conducted in Pichavaram, Tamil Nadu, Inglin (2013) investigates why the adoption of surukku valai fishing techniques suddenly emerged in post-tsunami coastal Tamil Nadu. He traces the development of surukku valai nets among the fishermen of Tamil Nadu after the 1990s and the ways in which it degraded the coastal environment and caused the depletion of marine resources, as well as having negative effects on the roles of fisherwomen, despite the fact that it financially benefits surukku valai fishers.

FAO categorizes purse seines as one of the fishing methods of small-scale fishery systems (FAO, 2022b). Whereas the Indian government classifies fishing vessels into three types based on their "power" capacity. Purse nets fall under the category of automated vessels and do not belong to small-scale fisheries systems (Gunakar et al., 2017). To underline a clear distinction, based on fishing villages' widespread fishing methods, this book categorizes "small" and "big"[9] fishing villages. The small fishing village refers to a village primarily involved in artisanal fishing methods. In comparison, the big fishing village refers to the village that possesses substantial fishers engaged in surukku valai fishing. On average, the fishing village with a population count of less than 2000 was chosen for this research. I adopted two methods to identify the approximate population of the fishing villages: (i) by reviewing the recent census report by CMFRI (CMFRI, 2012a, 2012b) and (ii) discussions with the local leaders.

The major key informants of this study included leaders of local institutions, senior fishermen, and fisherwomen who have been active participants in women's self-help organizations. Informal interviews with local fisheries stakeholders helped me understand how the local institutions of Nagapattinam district's fishing villages are organized and self-governed without the interference of external agencies/actors; it also gave insights into the social, cultural, and political elements of Pattinavar fishermen along the coastal stretch. Employees of a local NGO gave information about the social and geographical vulnerabilities of fishing communities in the Nagapattinam area. Based on a series of fieldwork activities and discussions with local key informants, I chose seven fishing villages to undertake this study: Kodiyampalayam, Koozhayar, Madavamedu, Kottaimedu, Chinnakottaimedu, and Chinnamedu & Chavadikuppam. According to Nagapattinam's disaster management department, all these villages are prone to coastal disasters.

Descriptive notes on the study villages

Kodiyampalayam: Kodiyampalayam is a small coastal village of Kollidam block in Sirkazhi taluk, Nagapattinam district, Tamil Nadu. It is on the border of Nagapattinam district and the nearby Cuddalore district of Tamil Nadu.

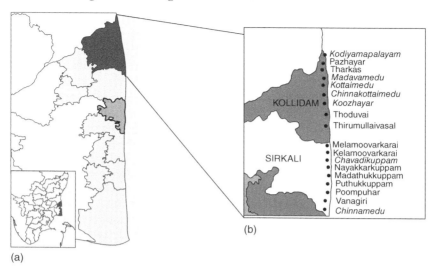

Figure 3.4 Locating the study villages (maps not to scale, only for illustrative purpose) (a) Wikipedia Commons (2015). https://commons.wikimedia.org/wiki/File:TN_Taluks_Nagapattinam.png (b) Study villages.

It is surrounded by water (Ocean and river) on all sides; hence, the geographical location of this village makes it appear like an island. A small bridge connects this village to the nearby villages. Locals call this village *Kodiyampalayam theevu*.[10] Discussions with the senior fishers show around 480 households in this village, of which around 450 families[11] are dependent on marine fishing. The approximate population of this village is 2000. About 20 families who are non-Pattinavar reside in this village for decades, and they have continued to engage in local occupations other than marine fishing. All the remaining households are Pattinavars, and they are entirely dependent on small-scale marine fishing by their FRP-OBM units.[12] Substantial households in this community do not own fishing vessels or nets; instead, they make a living as fishing laborers. Hence, even though more than 90% of the people are Pattinavars and Pattinavar uur panchayats operate the village entirely, it cannot be regarded as a single-caste fishing community in the strictest sense. In addition, a few males from non-Pattinavar families work as fish laborers, which allows them to participate in fishing. Fishing is done in groups, and in most cases, they collectively invest in fishing and share their profits and losses along the lines of their "formula." Around 150 men from the fishing households of this village are working in foreign countries[13] (mainly gulf countries, Singapore, and Malaysia) as manual laborers. In general, the economic status of most households in this village is inferior to that of the adjacent large fishing villages. This village was severely hit by the 2004 Indian Ocean Tsunami and the subsequent extreme weather conditions. Some foreign NGOs, including CASA, constructed approximately 150 free

Unraveling climate change and disasters 49

dwellings on the land of fishermen and donated them to them. The government of Tamil Nadu also helped a few fishing households build concrete houses. Nonetheless, local authorities claimed that the welfare benefits were insufficient. The majority of the fishing households in this village reside in concrete houses, and the remaining ones are in brick-roofed houses. The dwellings of most fishers are stable and can stand up to normal harsh weather conditions. In addition to marine fishing, fishermen often fish in the neighboring Kollidam River during periods of fish bans[14] and rough seasons.

Koozhayar: Located in the Nagapattinam district, the small fishing village of Koozhayar falls under the Vettangudi panchayat. Marine fishing is a primary source of income for nearly all households in this village. The total population of this village comes to around 1500. According to the respondents, the fishers of Koozhayar own 2 surukku valai boats, 25 OBM boats, and 30 FRP Catamarans. A group of fishermen in this village started surukku valai fishing a few years ago. The tsunami waves hit this village and caused extensive damage to the families of Koozhayar. The majority of fishing households in this village are the victims of tsunami disasters as well as they are the beneficiaries of tsunami recovery and rehabilitation programs. Like Kodiyampalayam, the humanitarian organization constructed tsunami shelters on the land of fishing families in Koozhayar and provided them for free.

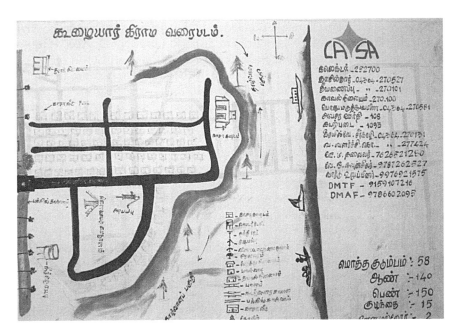

Figure 3.5 Village map of Koozhayar.

(Prepared by CASA Non-Governmental Organization during the tsunami relief and recovery.)
Place: Koozhayar. Picture credit: Devendraraj Madhanagopal.

According to the respondents, the residents live in two different regions of the village: (i) Tsunami-Nagar, where the free-tsunami householders reside.[15] Tsunami-Nagar is around 800–1000 meters away from the coast. (ii) The Main (Old) Village: It's roughly about a one-and-a-half-kilometer walk to the beach. Locals often refer to this area as the "old area" or "westward" because most of the fishing families in this village used to live close to the coast before the tsunami hit. Even though they had homes in the old area (westward), they liked living near the beach because it was easier for them to fish there. Besides that, they also wanted to keep outsiders from taking their fishing assets. In general, all of the fishing families in this village own houses.

Rent-homes are uncommon in the fishing villages along the coast. This village has a diverse group of fishers. Most of them are involved in small-scale fishing, and a significant number of them are fish laborers, and a few families are also engaged in surukku valai fishing.[16] Fishers of this village are collectively engaged in fishing, and they share the profits, losses, and risks among themselves based on their local formula.[17] In a strict sense, this village is not a homogeneous single-caste Pattinavar fishing village as a small proportion of Scheduled Caste families and families belonging to OBC also reside in this village. The norms and principles of Pattinvar uur panchayats, however, control this village. The socioeconomic situation of fishers as a whole is deteriorating. In Koozhayar, women's self-help groups are active, and they are keen to bring in outside/additional sources of revenue to their fishing households, albeit they are rarely successful.

Chinnakottaimedu: Chinnakottaimedu or Olakottaimedu is a small coastal village in Sirkazhi taluk, Nagapattinam district, Tamil Nadu. According to locals, this village is the smallest in the Nagapattinam district in terms of population. According to Census data, 75 of the 197 households in the study fall below the poverty line (CMFRI, 2012b). While the fieldwork revealed that around 250 fishing families live in this village, which has an estimated population of around 800 people. The tsunami caused devastating losses of life and means of subsistence for the fishers of this village, and they remain vulnerable to climate change and other risks. Most of this village's fishing houses were relocated to the newly constructed Tsunami-Nagar following the tsunami disaster. Compared to the other study villages, this village's infrastructure is in poor condition. Only about two-thirds of households are in concrete houses, while the rest are in brick-roof houses.

Kottaimedu: Kottaimedu is a small fishing village in Sirkazhi Taluk of Nagapattinam district, Tamil Nadu. According to CMFRI (2012b), the population of this village is 776,[18] and 196 families are involved in traditional small-scale fishing. According to the locals, this village has a population of around 1200 people, with 450 fishing households. There are 12 surukku vaalai boats, 110 FRP-OBM units, and 10 FRP Catamarans owned by the fishers in this village, according to local leaders. As with the other study villages, this village was severely affected by the 2004 Indian Ocean Tsunami. Fishermen in this village benefited greatly from tsunami relief and reconstruction efforts. Over 165 families in the village received free tsunami houses. Some fishing families own residences in coastal areas, such as in the vicinity of the beach and Tsunami-Nagar.

Unraveling climate change and disasters 51

Figure 3.6 A small fish market center in Chinnakottaimedu – Also known as Olaikottaimedu (by Red Een Kind as a part of the tsunami relief program).
Picture credit: Devendraraj Madhanagopal.

Figure 3.7 Fishing vessels and nets. Place: Chinnakottaimedu beach.
Picture credit: Devendraraj Madhanagopal.

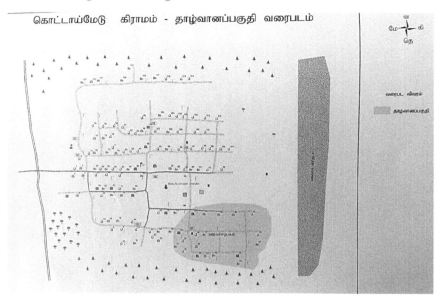

Figure 3.8 Low-lying areas in Kottaimedu.
Prepared by The Hunger Project.

Surukku valai fishing is prevalent among Kottaimedu fishermen. OBM boats are used to transport their fish catch from the sea to the beach. They are, however, reliant on OBM boats to catch fish during the fish ban and rough seasons. Overall, the fishing families in this village are better off financially than those in the nearby fishing hamlets, including the study villages. As part of tsunami rehabilitation efforts, humanitarian organizations supplied approximately 45 OBM boats to the fishers of this village. This village also received disaster management tools, such as rope and tires. The Tamil Nadu government's fisheries department recently gave free life jackets and rubber tubes to this village, which the fishermen found useful in many ways. Nonetheless, due to space limits in their fishing vessels, fishermen do not wear life jackets when fishing. During the fieldwork, I often noticed that fishers throughout the study region are casual about preventive measures, and Kottaimedu fishers are no exception.

Madavamedu: Madavamedu is a small fishing village in Kollidam taluk of Nagapattinam district with an approximate population size of 1200 with around 400 fishing households. This is a single-caste Pattinavar fishing village. As per the statements[19] of senior fishermen, Madavamedu was wreaked by the tsunami waves with 12 fatalities – most of them were children. Almost all the fishing households are the beneficiaries of tsunami rehabilitation measures. Fieldwork suggests to me that the overall economic condition of this village is relatively better than the other study villages (except Kottaimedu) as the majority of fishing households of this village are surukku valai fishers. Hence, most of the fishing households of this village reside in concrete houses, and a small proportion of them reside in brick roof houses.

Figure 3.9 Fish selling place – Built by IFAD as a part of tsunami rehabilitation measures. Place: Madavamedu.
Picture credit: Devendraraj Madhanagopal.

Chinnamedu: Chinnamedu is a small-scale fishing village/hamlet in the Nagapattinam district. It is about 40 km north of Nagapattinam, which is the district capital. The village's fishers are the front-line victims of the 2004 Tsunami, the 2011 Thane cyclone, and the 2015 heavy rains. They rely heavily on OBM boats and catamarans to sustain their fishing livelihoods in this village, and many of them are involved in the nearshore fishery. The total number of fishing households in the village comes to around 250, and the population comes to around 1000.[20] This village's houses begin just 200 meters from the seashore. Consequently, during rough seasons, the fishing households of this village are extremely at risk. This village is far more vulnerable to coastal erosion – most fishers have reported dwindling coastal spaces over the years. Moreover, the village doesn't have good drainage or sewage systems, which makes the people there more vulnerable during and after extreme weather. During every rough season, the rainwater remains stagnant in the village for several days. Concrete houses are common in Chinnamedu, but many of them are not built to withstand extreme weather conditions.

Chavadikuppam: Chavadikuppam is a single-caste Pattinavar fishing village in the Sirkazhi block of Nagapattinam district with a population of about 1000. Around 400 Pattinavar fishing households and 30 households belonging to the OBC castes reside in this village. The livelihoods of all the fishers of this village are entirely dependent on small-scale fishing, and none of them are surukku valai fishers. The tsunami disaster claimed the lives of 15 villagers (including children). As part of the tsunami relief and recovery efforts, around a hundred free houses were given to the fishers in this village. This village is divided into two parts (west and east). A substantial number of fishing families can be found west of the coast, where the free-tsunami dwellings are concentrated. It is about 700–800 meters from the coast. Even though

54 *Unraveling climate change and disasters*

Figure 3.10 Fishing vessels of Chinnamedu fishers (both FRP-OBM units and small non-motorized canoes).
Picture credit: Devendraraj Madhanagopal.

Figure 3.11 Fishing vessels. Place: Chavadikuppam beach.
Picture credit: Devendraraj Madhanagopal.

the village is quite close to the beach, some fishing families still live in the other part (east) of the hamlet (less than 400 meters). Chavadikuppam fishers are mostly concerned about the effects of coastal erosion.[21]

Summing up the study villages

The study villages are highly homogeneous, single-caste (Pattinavar) fishing villages, and most fishing households (except those in Madavamedu and Kottaimedu) rely heavily on nearshore small-scale marine fishing for their livelihoods. Around 35%[22] of the fishing households of the study villages do not own vessels and nets, and the fishermen from these households earn income by working as fish workers. Along the whole coast, fishing has always been a predominantly male occupation. As a result, the primary source of income for fishing households is primarily dependent on male heads. Significant senior women in their late forties and fifties from the study villages market the fish as head loaders. Major disasters that affected many fishers in this region include the 2004 Indian Ocean Tsunami, the Thane cyclone of 2011, and the massive rains of 2015. However, fishers reported receiving several welfare benefits from external actors in the aftermath of the 2004 tsunami; since then, they have been repeatedly victimized over the past several decades owing to a variety of climate extremes. Nonetheless, the welfare measures provided by the state and other external entities are woefully inadequate. Over the past two decades, the fishing community has seen a major economic shift, and the overall income patterns of fishermen have typically risen. Furthermore, the tsunami provided numerous opportunities to improve fishers' living standards and economic conditions (Fihl, 2014). So, it is hard to claim that most fishing households in the study villages are extremely poor. The fieldwork also suggests that the food security of Nagapattinam district fishers is not jeopardized. This is owing to the state government's extensive social welfare initiatives and financial assistance to marine fishers.

Qualitative approach

Mounting literature discusses the qualitative research methodologies and their applications in case study research (Walker, 1985; Strauss, 1987; Silverman, 2000; Yin, 2003). However, it is difficult to provide a single definition for qualitative research as it represents the umbrella of various strategies (Byrne, 2001). According to Shank (Shank, 2002, p. 5), qualitative research is *"a form of systematic empirical inquiry into meaning."* Stake (2010) highlights some of the basic qualities of qualitative research, such as interpretive, experiential, situational, personalistic, and so on. Qualitative research is a methodologically rigorous, labor-intensive, and time-consuming procedure. Starman's definition of qualitative research has reflected this view, in which he defines that "qualitative research characterizes by its interpretive paradigm, and it emphasizes the subjective meanings and experiences of the respondents and the researcher." The most common qualitative research methods are observation, interviewing, and examination of the documents (Starman, 2013). Qualitative research focuses on various social environments and the lives of those who inhabit such settings. It examines how humans have organized themselves socially through symbols, rituals, social roles, culture, and social structures (Berg, 2009). Qualitative research investigates human experiences, perspectives, and motives (Austin & Sutton, 2014). By highlighting the significance of qualitative research, Silverman (2000) argues that it should be given more consideration due to its "holistic treatment of phenomena." Yin (2011, pp. 7–8) distinguishes five features of qualitative research. The brief explanations of the features are as follows: (i) It studies the meaning of people's lives in a real-life context, (ii) it represents the views and perceptions of people, (iii) it investigates the phenomena in their contextual nature, (iv) it contributes to the concept of understanding complex social processes and behaviors, and (v) it uses multiple sources of evidence.

Rigorous fieldwork made me recognize that small-scale fisheries are highly complex and dynamic systems that involve a wide range of stakeholders – where everyone's roles and activities are delineated on the lines of caste, gender, class, and power. Hence, the social and institutional realities of marine fishers of not just in this research region but also in entire South India need to be understood in their natural settings. Fieldwork and the series of informal discussions with the key informants made me adopt a qualitative research approach as it justifies the demands of this research. This approach provides flexibility and freedom to conduct the research as well as to interpret and analyze the findings from the research (Eisenhardt, 1989). I conducted non-participant observations, informal talks, open interviews, semi-structured interviews, and focus group discussions with the fishers to collect the data. Also, I adopted the diary-writing method (Adamson, 2015) to note down the field notes and respondents' responses. All such interviews were held in the local language (Tamil) through proper schedules and checklists.

Data collection methods

Selection of the key informants

The key informants are the expert sources of information for this research (Marshall, 1996). Key informants have the opportunity to learn more about fishers, climate change, and governance systems because of their social status and fishing experience. This study requires fisher respondents with a strong understanding of the institutional arrangements of fishing villages and the past and present impacts of climate change. As a result, the selection of key informants was based on their age, fishing experience, designation, expertise, and knowledge of the changes in their local environment due to climate change. In most cases, senior fishermen – those with more than 25–45 years of fishing experience – served as key informants. Most of the key informants of this study were uur panchayat leaders, local body leaders, senior fishermen, representatives of women's self-help groups, and the fishermen cooperative society, and NGO employees. Along with them, I also interviewed a few academicians based in Chennai, Tamil Nadu, and a writer and an activist[23] from Kanyakumari, the southernmost end of Tamil Nadu. When selecting the samples, I initially sought to achieve a balance between male and female fishers. Whereas after the pilot survey, I recognized that the coastal villages exclusively govern the lines of caste and gender. Hence, the fishermen historically occupy key positions across all local institutions along the coasts, and the roles of the fishermen are largely confined to their households' activities; in some cases, it extends to earning income for their families by selling the fish as head-loaders. Due to this, fisherwomen are largely unaware of the existing and emerging livelihood issues in the coastal villages. Like this, their awareness of the changing climate is also limited. After several days of deliberation, I concluded that ensuring proper representation along the lines of social groups and gender will be of little benefit and may not meet the objectives of this research. Also, adopting a formal and structured questionnaire to elicit information from the respondents will not be enough to meet the demands of this study. Hence, to discuss climate change impacts, indigenous knowledge of fishers, and the local governance of the coastal villages, I mostly relied on the senior fishermen and the male leaders of the local institutions of the study villages. To ensure the representation of the women in this research, fisherwomen who are actively engaged with the women's self-help groups were also included in the interviews and focus group discussions. Proper schedules and checklists guided in-depth open interviews and focus group discussions. I attempted to explore the documents of the local institutions to understand how the fishing villages managed to respond to climate change (Adamson, 2015). I, however, was informed that the local institutions do not systematically document such details and preserve the records to date. In the earliest days of fieldwork, I attempted to elicit the observations, opinions, and experiences of fishers regarding climate change and dangers over the previous several decades by eliciting their indigenous knowledge of reacting to climate change through informal conversations, unstructured open interviews, and focus groups. Aside from the main players in the local institutions, fishermen, fisherwomen, and political party representatives were also interviewed to get a full and coherent picture of how the local institutions work together. A special emphasis was placed on understanding the wide consequences of climate change on fisher livelihoods as well as the ways in which the key actors of local institutions approach climate change threats.

Sampling strategy

Purposive sampling is often used by qualitative researchers to choose respondents instead of the rigid, traditional method of random sampling. Purposeful sampling allows the researcher to find respondents who can provide detailed insights and rich data to meet the requirements of the study (Hanson et al., 2011). Snowball sampling is a subtype of purposive sampling. Here, the researcher asks respondents to identify specialists or additional participants who could contribute valuable information for the study's research aims (Patton, 2002; Hanson et al., 2011). Rather than "generalizability," the objective of qualitative research is to glean deeper insights and comprehensions about the case (Hanson et al., 2011; Palinkas et al., 2015;

Table 3.3 Details of the respondents

Respondents	Methods of data collection	Numbers of respondents
Fishermen	In-depth interviews and focus group discussions	150 (including 40 key informants)
Fisherwomen (majority of them are active in women self-help groups[1] and head loaders)	In-depth interviews and focus group discussions	40 (including 10 key informants)
Middlemen (fishing)	Informal discussions	10
Fish traders (all are Pattinavars)	Informal discussions	10
Retired fishermen	Informal discussions	20
NGO employees (IFAD, MSSRF, ICSF, SIFFS)	Informal discussions (mostly during the pilot study)	Around 10 (only 4 respondents participated in the discussions. Remaining respondents helped identify key fishing village informants)
Local businessmen (other than fishing)	Informal discussions	4
Government employees	Brief informal discussions	3
Activists and writer	Discussions including e-mail discussions	3
Academicians	Discussions conducted before pilot surveys	3

1 Throughout the book, I use the terms "women self-help groups" and "women's self-help groups" interchangeably.

Rosenthal, 2016). It requires the researcher to focus on the study's research topics and to maintain dialogues with respondents. The modest number of samples used in qualitative research is justified by doing an in-depth investigation and gaining new knowledge (Crouch & McKenzie, 2006). Purposive and snowball sampling approaches were employed to locate and choose the participants in this study.

Data analysis and methods adopted: some reflections

From May 2014 to February 2018, I conducted five stages of extensive field trips to collect data across different time intervals (including the initial pilot survey). To collect the data, I stayed in the field and interacted directly with the respondents. Around 20 days were spent discussing with field experts and

Table 3.4 Age group and fishing experience of the fisher respondents

Category	Number of Respondents	Age group (in years)	Fishing experience (in years)
Senior fishermen	80	46–65	25–45
Middle-aged fishermen	40	36–45	15–25
Young fishermen	30	25–35	10–15
Young and middle-aged fisherwomen	35	25–45	0–10 (fish selling experience and experience with women's self-help groups)
Senior fisherwomen	5	46–65	10–20 (fish selling experience and experience with women's self-help groups)

* Only fishermen and fisherwomen, including the key informants, are included.

academics to have a better knowledge of Pattinavar fishermen's lives, livelihoods, vulnerabilities, and governance systems. Overall, I spent about four and a half months getting all the information I needed for this study. After getting oral consent from the respondents, I recorded substantial interviews and focus group discussions. Few fishermen expressed reluctance to discuss their village's problems because they feared it would intensify them in the future. In such circumstances, I recognized the scenario and chose not to record their responses. I recorded the fishermen's responses throughout the data-gathering process in diaries, notebooks, and separate data sheets (in some instances). I highlight some selected accounts from the field in the forthcoming chapters of this book, but I do not divulge their identities to protect the respondents' confidentiality. Instead, I highlight many identifications of the responders to ensure the evidence and legitimacy of the conclusions. The following are some of them: (i) the name of the respondents' village, (ii) designation of respondents in fishing villages, and (iii) fishing experience of respondents. In qualitative research and analysis, the researcher should concern the two important factors: validity and reliability (Patton, 2002; Berg, 2009). In research, validity is concerned with the accuracy of scientific findings (Le Comple & Goetz, 1982) and reliability is concerned with the consistency, stability, and repeatability of the informant's accounts as well as the investigators' ability to collect and record information accurately (Selltiz et al. 1976, cited in Brink, 1993). I took several steps to ensure the validity and reliability of the data. I collected the data by orally translating the questionnaires and checklists, described to the respondents, and noting down their responses in diaries and notebooks. I did the entire data collection without anyone's assistance.

Furthermore, during my early visits, I learned that semi-structured interviews were only useful for understanding the socioeconomic conditions of the fishermen, but not for eliciting information about climate change and local government systems. Before collecting any data, the purpose of the study was always communicated to the respondents and their oral consent was obtained. I conducted most interviews and focus group discussions with the aid of checklists and recorded their replies in English (and sometimes, Tamil) by using notebooks and audio equipment.[24] I gave the respondents wide rein to express their views on climate change, its effects, internal governance structures, and the responses of local institutions. Respondents were sometimes unwilling to divulge information, particularly about their boats, money, etc. They were also hesitant to give specifics about certain sensitive matters. I understood and adapted the interviews based on the respondents' convenience. The quality of my interviews and focus groups was not compromised, though. In addition, I took care, as recommended by Islam (2013), to end interviews and focus groups when they reached saturation. Small-scale fishing communities face a wide range of challenges, and climate change and extremes are not the only factors at play. Some respondents voluntarily shared information about overfishing and highlighted their grievances toward mechanized fishers. In such instances, after reaching specific boundaries, I halted the interviews and focus group discussions on purpose and attempted to concentrate on the flow of the interviews/discussions in order to achieve the aims of this study. Before beginning the data collection process, I met with local leaders, had brief talks (mostly informal discussions), and collected information on their responses. During the series of interviews and focus group discussions at different time intervals, I also tried to meet with similar respondents and conducted the same interview I had done earlier. I then took notes on their responses and cross-checked them against the prior responses to ensure their correctness. It guaranteed the data's dependability. I followed the descriptive qualitative analysis method (Creswell, 2014) to analyze the data. My questionnaires and checklists were in English. Besides, as I noted earlier, I recorded respondents' responses. It was helpful to me to have a detailed account of the information. Firstly, I translated all the recorded interviews and focus group discussions from Tamil to English and categorized them based on their themes using excel sheets. Finally, the findings were manually interpreted.

Few challenges of the fieldwork and clarifications

It took me a lot of time and effort to select and locate respondents who would be willing to provide the level of detail necessary to achieve the study's aims. Liquor consumption is a chronic issue among Tamil Nadu's fishermen. During the fieldwork of this study, I found that most fishermen take liquor after a typical fishing day. Consequently, gaining access to suitable respondents, including the key informants, was challenging. I turned to the village heads for assistance in resolving this issue. Every day, from noon to evening, I would go to the beach and look for a few fishermen (mostly senior fishermen) who would chit-chat with their colleagues (native villagers) and repair their fishing nets and vessels. Therefore, selecting potential senior fishermen and fisherwomen respondents for this study was an arduous experience. In Tamil Nadu, fishermen's panchayats are run solely by men, and as a result, the state's internal governance structures are dominated by fishermen. Fisherwomen mainly engage in domestic duties, and only a small proportion actively participate in women's self-help groups. Active women self-help group members of the villages were eager to participate as a respondent in this study, whereas the remaining fisherwomen showed hesitancy. Besides that, it was hard to reach the fisherwomen directly because it could cause other complications. Because of this, I attempted to reach out to local IFAD employees, who referred the secretary of IFAD women's self-help groups. After going through these challenges, I chose appropriate women respondents with the help of self-help group members. Nevertheless, a limitation of this study is still the fact that there are significantly fewer "female" respondents than "male" respondents. Communities that rely upon natural resources for their everyday livelihoods are aware of the small changes as they continuously observe and locate the differences in their weather and local environment (Orlove et al., 2010). Though their climate perceptions do not have scientific validity and are not on par with scientific methodologies, it has their value in understanding the broad impacts of climate change on their everyday livelihoods. Likewise, Pattinavars' perspectives on climate change do not come from scientific understanding; instead, it is the result of their rich traditional ecological and experiential knowledge of fishing and everyday experiences with the sea and the coast. Therefore, I also had to alter and modify the schedules and discussions according to the participants by considering their age, fishing experience, and expertise to provide me with the required details. Olsson et al. (2014) describe how climate change interacts with other non-climatic stressors and intensifies the vulnerabilities of the poor. They use "weather events and extremes" as an "umbrella term" to denote the broad impacts of climate change and variability. I followed Olsson et al. (2014) and applied the term "climate change" as an "umbrella term"[25] "that encompasses global climate change associated with weather and climate variability, including climate extremes, and gathered the primary data from the fishers.

Notes

1 Envis report (n.d.) provides an outline of the geography, weather and climate patterns, development, and environmental vulnerabilities of the Nagapattinam district.
2 These villages are significantly more populous than the neighboring small fishing villages. Because of their population size and economic standing, these villages have received more attention from the state and political parties. Thus, the infrastructure and other amenities of these villages are considerably superior to those of Kodiyampalayam and Koozhayar, which are located nearby.
3 This chapter contrasts "big" and "small" fishing villages, following a summary of the study villages. In essence, "big" fishing villages are those that are predominantly engaged in mechanized (large-scale trawler) fishing, whereas "small" fishing villages are those that are predominantly engaged in small-scale fishing (artisanal fishing methods, FRP boats).
4 Marine fishers are generally at the lower socioeconomic strata throughout India, and Tamil Nadu is no exception (Bavinck & Vivekanandan, 2017). In Tamil Nadu, fishing groups are considered "Most Backward Castes" and benefit from the national and state reservation systems (quotas).
5 Here, all the italicized words denote the caste names.

6 Pattinavar fishermen in some places also refer to themselves as "Chettiar fishermen." However, officially and commonly, fishers are referred to as Pattinavars and Meenavar Pattinavars (Fihl, 2009).
7 Figures 3.1, 3.2, and 3.8 were obtained from the Hunger Project reports that were on hand in the corresponding local panchayat offices. This book only referred to hard copies of reports accessible at the panchayat offices. The book's reference section includes the Hunger Project India's website information.
8 All the fishermen of the study villages fish with engineless FRP boats and FRP-OBM fishing vessels. Madavamedu and Kottaimedu fishers and a few Koozhayar fishers engaged in surukku valai fishing. In English, Surukku valai fishing is called Purse-seine fishing. Local fishermen refer to this type of net as a "purse net" or a "ring net." Senior fishermen also explained that "bottom trawlers" and "purse net" fishermen are vastly different in their methods. While some local heads of Madavamedu and Kottaimedu stated that the mesh sizes of their nets are permissible, these are oral assertions without conclusive proof. To ensure clarity and avoid ambiguity, I prefer to use the Tamil term "surukku valai" to refer to "Purse net/Ring nets" fishing methods of the fishers. In addition, as will be described in the following sections, I conducted fieldwork in the study villages between 2015 and 2018 in different phases. Surukku valai was not prohibited by the government when I conducted fieldwork in Tamil Nadu. An amendment to the Tamil Nadu Marine Fishing Regulation Rules, 1983, which went into effect in February 2020, prohibits the use of purse seine nets – surukku valai (Government of Tamil Nadu, 2020). It sparked widespread unrest among marine fishers in the district of Nagapattinam, including fisherfolk from Madavamedu. According to local news sources, they participated in great numbers in protests against the Tamil Nadu Fishing Regulation Act (The New Indian Express, 2021). Fisheries and Fishermen Welfare Policy Note (2022) for the year 2021–22 also clearly prohibits the use of purse seine by any type of fishing vessel in Tamil Nadu's coastal waters (Government of Tamil Nadu, 2021). Whereas the government of Tamil Nadu's most recent Fisheries and Fishermen Welfare Policy Note 2022–2023 is silent on the ban on surukku valai fishing (*Government of Tamil Nadu*, 2022a). These are the most recent occurrences. Whereas the arguments and discussions in this chapter are based on the fieldwork data that was collected.
9 Hereafter, the usage of inverted commas will be avoided to denote small and big fishing villages ensure the flow of reading.
10 In Tamil, it means Kodiyampalayam Island.
11 The populations, fishing households, and other quantitative facts about the study villages given in these parts are based on oral accounts from local leaders and senior fishermen. Hence, it differs from the official statistics.
12 Three types of fishing vessels are commonly owned by fisherfolk in the sampled villages. (i) Catamarans. (ii) Country crafts, including canoes (vallam: in Tamil). (iii) Maruti FRP boats: Maruti boat is a small boatyard. Fishers use outboard motorboats (OBM boats) that are easy to fit on catamarans and Maruti model boats. Small light diesel motors with a long shaft and a propeller provide power to drive the boat. Long-tail OBM boats are meant for small-scale fishing. To learn more about the artisanal and mechanized fishing units operating along the coast of Tamil Nadu, check out FIMSUL (2011). This publication provides a wealth of information about the fishing methods, gear, and vessels currently in use by Tamil fishermen. It also describes how FRP boat fishermen utilize ring-seine units (surukku valai fishing) as a seasonal alternative. Like this, TNAU Agritech Portal (2015) provides an account of fishing crafts of the fishers of Tamil Nadu with images.
13 The migration of fishermen to foreign nations is prevalent along the Tamil Nadu coast. Therefore, in the area of Nagapattinam, it is normal to encounter numerous fishermen with working experience in foreign nations. Significant fishing households in the study villages have at least one family member working abroad

as a "manual laborer/fish worker." If not, they used to spend years working in foreign countries before returning to their native villages to engage in their traditional occupation, fishing. Lack of money, declining catches, and other economic factors are the primary reasons why fishermen leave. In recent decades, climate change has forced fishermen to leave their villages. Remittances from migrant fishermen help their families cope with stressors.

14 In 2017, the state government of Tamil Nadu decided to extend the fishing ban for mechanized boats from 45 days to 61 days (i.e., 15 April to 14 June) to promote fish reproduction in the oceans. The extension period and timings also received critiques from fishers (Badola, 2020).

15 Many houses in Tsunami-Nagar are not in proper condition to withstand climate risks. Fishers do not have any regular income sources other than marine fishing. Hence, they could not reconstruct their old free-tsunami houses.

16 The majority of Koozhayar fishers are involved in small-scale nearshore fishing. The harvest from one surukku valai net is shared by around 50 fishermen. The remaining fishermen use OBM vessels to catch Seer/Kingfish (Vanjiram, in Tamil) and other fish species.

17 Surukku valai fishermen share collective ownership of fishing assets and catch. This formula varies from village to village among fishermen. Fishermen's roles and investments in fishing vessels tend to determine how much they share.

18 The fieldwork for this research provides different data. This village is a single-caste fishing village with a population of around 1000. Fieldwork revealed that villagers own 110 OBM boats and 12 surukku valai boats. Fishermen call this boat "*visai padaku.*"

19 Around 292 fishing households received free tsunami houses from donor agencies. Fishers locally refer to the surukku valai fishing boats as "steel boats." The steel boat ownership is collectively shared among the fishers. More than 60 fishermen from Madavamedu collectively own a single steel boat. This number varies according to the overall economic status and population of the village. They further noted that the surukku valai fishing method is restricted during the fish ban period (May–June, 45 days). Uur panchayats informally monitor and regulate the fishing activities of fishers during this period to avoid conflicts between the fishing villages and the state.

20 Senior fishermen and local leaders said around 80 adult males of this village work abroad as manual workers. They stated that around 145 fishing households live in free-tsunami houses. They were also worried about how well the houses could stand up to the growing risks of climate change.

21 The state government recently announced its plan to construct seawalls on the coasts of Chavadikuppam along with the other nearby fishing villages. However, the process was pending (as per the fieldwork from 2015 to 2018).

22 This percentage is approximate for the following reasons. During the field visits, few respondents were not interested in disclosing the precise ownership information of their fishing nets and boats. Few fishermen own non-motorized catamaran boats but prefer to work as fishermen because fishing is an "uncertain" job. In some cases, fishermen own just fishing gear, not fishing vessels.

23 Chris Antony: He is a writer from the Kanyakumari district. *Thuraivan* is his Tamil novel, and he blogs about the lives, culture, and rights of fishers. I communicated with him via email in order to gain a better understanding of the difficulties faced by fishermen.
Vijayan: He has been a fishers' rights activist for almost two decades. He is a writer from the Kanyakumari district. I had casual conversations with him to learn about livelihood and coastal erosion. The writers listed here are from Kanyakumari, not the region where I conducted my research. However, my conversations with them have provided me with a greater understanding of the modern obstacles faced by Tamil small-scale fisherfolk.

24 Approximately 50% of respondents readily consented to the recording of their responses. The audio was then recorded. Social and economic data were gathered through semi-structured interviews, and the study's core data, which related to its research objectives, were obtained primarily through in-depth, open interviews, primarily in the form of fishermen's personal narratives. Like this, all the images of the fisher respondents that I share in this book were taken with their permission. They gave me their verbal permission to take their photos and use the research in various ways.
25 Chapter 4 on indigenous knowledge systems provides a detailed understanding of this.

4 Indigenous knowledge systems in confronting climate change
Opportunities and constraints[1]

Local/indigenous knowledge systems: an overview

Multiple global studies indicate the importance of local perspectives and local knowledge in climate change research and disaster risk reduction (Vedwan, 2006; Shaw et al., 2009; Mercer et al., 2010; Kelman et al., 2009; Kelman, 2010; Kelman et al., 2012; McNamara & Westoby, 2011; McNamara & Prasad, 2014; Nehren et al., 2013; Macchi et al., 2015; Musinguzi et al., 2016; Guerreiro et al., 2017). There has been a growing scholarly and research interest in climate change perceptions of natural resource-dependent communities in South Asia in recent years (Halder et al., 2012; Piya et al., 2012; Mathur, 2015; Shukla et al., 2016; Roy, 2020). Local knowledge advanced by academics can be combined with scientific knowledge (Becken et al., 2013; Hiwasaki et al., 2014; Ford et al., 2016). However, traditional knowledge-based methods have had strong critics (Widdowson & Howard, 2008). It is a misconception, according to Berkes (2008), that all traditional/indigenous practices are environmentally sound and sustainable. Tibby et al. (2007) state that local/indigenous knowledge of environmental change should be validated by other sources, especially science, lest it skews our perspective of our local environment. As Byg and Salick (2009) noted, local observations and interpretations of climate change are not equivalent to scientific understandings of climate change due to their limitations. However, it might be useful as a supplementary scientific resource for scientists and policymakers. Knowing the social, cultural, and moral components of the local community is particularly needed to understand how communities deal with climate change locally. It is difficult to draw clear lines between indigenous and scientific knowledge due to their overlap in some ways (epistemological and practical), as emphasized by Agrawal (1995). He discusses multiple knowledge domains, each with its own logic and epistemology, and raises questions about the strict differentiation, delineation, and dichotomization between scientific and indigenous knowledge that is advocated for productive dialogues between the knowledge systems. Mazzocchi (2006) advocates the pluralist approach to understanding knowledge systems by acknowledging their uniqueness and historical contexts. Mistry and Berardi (2016) express a similar point of view, arguing that expanding the range of actionable indigenous knowledge

DOI: 10.4324/9781003187691-4

options will greatly benefit indigenous communities without attempting to validate their indigenous knowledge systems. As suggested by Singh et al. (2017), this understanding of local perceptions will serve as a supplement for assessing the vulnerability of communities and regions to climate change in various epistemic frames. As their livelihoods are intertwined with the marine environment, local populations in coastal areas experience climate-related changes daily. Consequently, it is necessary to record their perceptions and local knowledge and link them to adaptation planning for climate change impacts (Aswani et al., 2015).

Indigenous knowledge: an overview of Pattinavar fishermen's community lingo

Pattinavar fishermen have traditional names/community lingo to demonstrate winds, sea currents, and seasons in Tamil. They take into account the speed and direction of the winds and sea currents, as well as short-term predictions on fish quantity, before setting out to sea for fishing. Since Pattinavar fishermen have ventured into the sea for generations and centuries, a large portion of their community lingo is used by fishermen, but not fisherwomen. In addition, as discussed in the preceding chapters of this book, Pattinavar fisherwomen are primarily confined to post-harvest activities and domestic chores, and other family responsibilities. Pilot study results, therefore, suggested that most fisherwomen still lack the knowledge to provide accurate information on the shifting weather and climate patterns in their local environment over the past decades despite being aware of the ongoing changes. Given this, interviews and discussions related to the community lingo of Pattinavars that this chapter presents are confined to fishermen.

In the pilot survey, it was found that fishermen's indigenous knowledge of climate elements, oceanographic factors, fishing, and marine ecosystems is influenced by factors such as age groups, fishing experience, and the types of fishing vessels they use. According to Santha (2008), indigenous knowledge of fishermen varies depending on the type of gear used, family structure, and geographical location. Like this, as noted by Patt and Schröter (2008), the opinions of locals on climate change events are influenced by a variety of factors, including their memories. Fishermen's local observations of the effects of climate events were frequently not pinpointed. Age, education, and the types of fishing gear were the most influential factors in determining the fishermen's biases. This study backs up Santha (2008) and Patt and Schröter (2008). Also, the age groups of the fishermen play a significant role in determining the baseline of their perceptions regarding the effects of climate change. Local observations of young and middle-aged fishermen on climate change were frequently inconsistent. Only senior fishermen with at least 25–30 years of fishing experience were able to elaborate on climate change impacts and their various manifestations on species. To secure Pattinavars' traditional baseline of knowledge, respondents were preferred based on their age, fishing experience, and local knowledge about fishing practices and environmental changes. Those with more than 25 years of continuous fishing experience were given priority. Aside from that, fishermen in some villages, such as Kodiyampalayam, expressed a greater willingness to share their knowledge of fishing practices and climate change perceptions. In contrast, fishermen of Kottaimedu exhibited little interest in sharing their indigenous knowledge systems because they are predominantly surukku valai fishermen and rely more on scientific information for fishing. In selecting respondents, fishing experience, age, fishing type, and interests were considered. Further, to explore indigenous knowledge systems and climate change perceptions of Pattinavars, I did not adopt a similar sampling size across the fishing villages. This chapter presents interviews, focus group discussions, and non-participant observations with 100 fishermen from study sites. Olsson et al. (2014) describe how climate change interacts with non-climatic stresses to increase the vulnerability of the poor. They used "weather events and extremes" as an "umbrella term" to refer to the wide-ranging effects of climate change and variability. Using a similar method, I gathered data from fishermen using "climate change" as an "umbrella term" that encompasses global climate change associated with weather and climate variability, including climate extremes. According to Nakashima and Roue (2002), the use of the term "indigenous" lacks clarity because any knowledge can be labeled "local." Hence, to improve clarity, this book uses the term "local knowledge" to encompass the "local ecological" and "indigenous" knowledge systems of Pattinavar marine fishers throughout the region. On a typical fishing day, Pattinavar fishermen leave their homes at 4 a.m. to sea for fishing and return at 8 a.m. after selling the fish they have caught. It is the routine of almost all the fishermen across the Coromandel coast of Tamil Nadu. This is referred to by them as *"meenpaadu."* They (especially, small-scale fishermen of Kodiyampalayam, Chinnamedu, and Chavadikuppam) often do a few more fishing trips in a day. Fishermen plan their fishing activities according to the speed and direction of the winds and sea currents, as well as short-term forecasts of fish abundance, and then begin fishing in the sea. But such techniques and traditional names are largely only known by the senior fishermen. Young fishermen rely heavily on senior fishermen and scientific weather forecasts to predict weather and climate patterns.

Table 4.1 A few selected questions/schedules that guided the data collection

- Share the different traditional names (community lingo) of the sea currents, wind patterns, and seasons.
- How do you find out the timings, directions, and fish shoals when you are in the sea?
- What is the significance of wind patterns in detecting the fish shoals and determining the fish caught on a typical fishing day?
- How do the currents influence the fish catch and safely navigating the boats?
- How do you predict the impending rainfall, cyclones, and floods using local traditional knowledge?
- What are the signs and indicators of heavy rains, cyclones, and floods? Explain.
- How does the lunar cycle influence the fish catch?

(Continued)

Table 4.1 (Continued)

- Explain the connection between changes in climate patterns and fish catch.
- What are the climate variables that affect the fish availability and catch? List and rank the variables according to their influence.
- Explain the effects of the climate events that have occurred over the past three to four decades and point out how they have changed the fish-catching patterns over the years.
- Elaborate on the effects of the 2004 Indian Ocean Tsunami disaster on weather and climate patterns and fishing livelihoods.
- How did the 2004 Indian Ocean Tsunami disaster influence the changes in weather and climate patterns, and what are the changes in fish catch and fish stock in the aftermath of the tsunami disaster?
- Sources of weather-related updates and early disaster warning and its efficiency: fishermen's opinions and narratives.
- What are your views on the relevance of local traditional knowledge in the face of climate change? Is it beneficial to face and mitigate weather-related challenges while earning a sufficient income for your family?

Senior fishermen elaborated on how the effects of climate change over the past three to four decades have affected their livelihoods and the distribution and abundance of fish in their traditional fishing grounds. Fishermen use traditional names or community lingo[2] to describe the winds, sea currents, and seasons in Tamil. Based on their indigenous knowledge, fishermen classify the various types of winds. Fishermen on Tamil Nadu's Coromandel Coast have classified wind patterns into eight types. (i) *Naervaadai kaatru* (*Vaadai kaatru/Nedun kaatru/Vadamara kaatru*): This wind blows from north to south and occurs between November and February. (ii) *Naerchola kaatru*: It blows from south to north direction. (iii) *Naerkachaan kaatru*: It blows from west to east direction, and it happens in April, May, and June. Fishermen consider this wind as a "favorable" one as it eases them to reach the shore safely. (iv) *Naerkondal kaatru*: It blows from east to west direction. Fishermen consider this wind as a "rare one," but it may happen at any time throughout the year, except a few monsoon months such as October, November, and December. (v) *Vaadai kachaan kaatru/ Kunnu vaadai kattru/Saaral kaatru*: It blows from northwest to southeast direction, and it happens around November, December, and January. It has high velocity, and hence the fishermen consider it "unfavorable" to their fishing activities. (vi) *Vaadai kondal kaatru*: It blows from northeast to southwest direction, and it happens in October, November, December, and January. Sea gets turbulent during this period. Fishermen considered this wind as "unfavorable" to their fishing activities. (vii) *Chola kachaan kaatru/ Kachaan konda kaatru*: It blows from southeast to northwest direction, and it gets fierce from April to June. Fishermen get significant fish catch during this period. (viii) *Cholakonda kaatru*: It blows from southwest to northeast direction, and it usually happens from April to August. It gets more aggressive in April, May, and June (Samas, 2015; Primary data from the field). The nature of the sea currents varies according to the types of winds. In the field visits, it was found that the fishers broadly divided the seasons into two types: (i) *Kachaan* (Summer): From March/April to September and (ii) *Vaadai* (Rough season)[3]: From October to February/March. Fishermen across my study sites traditionally divide the sea currents into four types: (i) *Vanni vellam: Vanni vellam* and other types of *vanni vellam* come during *Vaadai* (winter). It flows from the north to south direction. (ii) *Soni vellam: Soni vellam* and other types of *soni vellam* typically happen during *kachaan* (summer). It flows from the south to the north. Fishermen consider this current favorable for fishing. It happens from April to August. (iii) *Memari vellam*: It flows from west to east direction. (iv) *Karaieduppu vellam*: It flows from east to west direction.

Environmental precursors that signal less fish catch

To foretell the start of heavy rain, experienced fishermen track the motion of the clouds. They say "*mappu*" to express the cloudiness of the weather and the impending rain. In general, fishermen prefer to fish during the summer season (except for April and May, when there is a fish ban) rather than during the rough season (October to December). Fishermen avoid taking their boats out more than 20 nautical miles during the rough season, according to observations made throughout the study area. Additionally, the early warning information (science) and local knowledge systems are crucial to their fishing expeditions during the harsh seasons. Senior fishermen generally study and analyze cloud movements to determine whether it will rain or not before traveling into the sea for fishing and giving guidance to local fishermen. According to the fishermen, if a cloud's color is black and it gathers in one location, it indicates the start of rain; however, if a cloud's color is white and it travels quickly with dispersion, it indicates a lesser or absent chance of rain. This is shown in Table 4.2.

The illustrations of fishermen's local knowledge that are shown in Table 4.2 largely concur with the observations of Swathi Lekshmi et al. (2013) who studied the indigenous technical expertise of marine fishermen in eight Indian coastal states, including Tamil Nadu. Although the sea becomes quiet after climate events, over half of the fishermen who participated in the study said they would not go fishing for at least two days. They assume that the situation will return to normal after two days and so on, but not immediately after the occurrence of the extreme situations. Senior fishermen recounted their stories of how they began heading into the sea for fishing in the aftermath of the tsunami calamity, despite just taking a four-day break. Most fishermen (including senior fishermen) were unaware of the migratory patterns of the fish species they often catch. As an illustration, they were asked to respond to questions about the migration patterns of Indian oil sardines, Indian mackerel, and flying fish that they typically catch in their fishing grounds and the causes of the migration patterns to look at their local knowledge systems. As shown by their answers, the fishermen clearly know very little about the migration and movement patterns of the species in question. A senior fisherman in Kodiyampalayam made the following remark:

Table 4.2 A brief illustration of fishermen's local traditional knowledge

Signs of high and low fish catch/fish shoals	• During the new moon period (*Amavasai*, in Tamil), more fish catches are obtained. During the full period (*Pournami*, in Tamil), fewer fish catches are obtained. • Flocks of birds' hover on the seawater indicate the fish shoals. • Coastal upwelling is an indication of good fish catch. • Dark blue patches, the presence of frequent bubbles and the ripples on the water indicate the fish shoals. • Presence of fishy odor at sea denotes the substantial fish catch. Whereas the bad odor at sea denotes the less concentration of fish in the fishing grounds. • Presence of muddy water indicates the substantial concentration of fish shoals. Whereas the clear white water suggests the less fish availability at the fishing grounds.
Signs of weather extremes including heavy rainfalls/ cyclones	• Sea water remains unusually very calm • Dark clouds are seen on the horizon. • The speed and intensity of the sea currents increases and the navigation of boats have become difficult because of the seawater. • Abnormal behavior of the animals (mainly dogs) and birds. • Foaming of water at the seashore is an indication of the impending cyclone.

Before two decades, we used to catch flying fish after making travels of about five "*thalai vaaram*" (In Tamil: One "*thalai vaaram*" denotes approximately 20 km. *Source*: field survey). But aftermath the 2004 tsunami disaster, we notice that we could catch flying fish by just making travels of around two "*thalai vaaram*." fishermen, aftermath earlier decades.

This study reveals that fishermen's ability to recognize fish shoals and forecast incoming rains/cyclones varies with age and years of professional experience. The growing severity and unpredictability of weather patterns in recent years, as described by fishermen, has made their occupation challenging. They often lamented that the irregular weather patterns made it impossible to detect fish shoals in the sea, which directly impacted their livelihoods and income. Since the 2004 Indian Ocean Tsunami disaster, the economic condition of fishing households across the coastal regions of Tamil Nadu has improved, leading to increased availability of information and accessibility to technology. This, in turn, has changed the behavior and interests of young fishermen, who have shown little or no interest in acquiring and comprehending their own local traditional knowledge from the senior fishermen. Thus, it is often observed that the transfer of local ecological knowledge from older

Indigenous knowledge systems in confronting climate change 71

to younger fishers has not been effective and that today's fishermen are more likely to use modern media sources than their elders for weather-related updates. Overfishing and climate change effects were cited by fishermen as the primary causes of the diminished and insufficient fish catch over the past two decades, especially after the tsunami disaster. The replies of all fishermen were completely symmetric, that is, overfishing of trawlers has decreased their fish capture over the last two to three decades. Along with this, the senior fishermen were of the same opinion that climate change is a reality and has directly affected the decrease in fish catch.

As demonstrated in Table 4.3, regardless of age differences, most fishermen largely noted changes in weather and climate patterns over the years, and denial of climate change was negligibly lower. Fishermen asserted that climate fluctuations, such as the increase in summertime temperatures and the worsening droughts over time, are the primary causes of species reduction and the diminishing fish catch over the decades. The extreme climate events and unstable weather patterns have gotten more intense and frequent, particularly following the 2004 tsunami, according to senior fishermen. As indicated in Table 4.3, most fishermen did not attribute the decline in fish catches solely to changes in weather and climate patterns over the decades. Instead, they thought about how climate change and overfishing had made it harder for them to make a living over time. According to them, these patterns have only escalated as the consequences of climate change have gotten more severe. Almost all the fishermen who took part in the study were worried about how they would make a living in the future because extreme weather was getting worse and happening more often. Most fishermen interviewed regarded climate change as one of the two principal causes of the continuing fish decline and the accompanying lower fish harvest, while the remaining senior fishermen claimed over-exploitation as the primary cause of the decades-long reduction in fish capture. It should be noted, however, that these elder respondents did not dispute that the weather had altered over time (Table 4.3). In their case study on cyclone warning information and public response on the east coast of India, Sharma et al. (2009) distinguish the terms "environmental cues" and "environmental precursors." They show how the knowledge and decisions of village elders and local leaders are the most important parts of the evacuation process when the state sends out a cyclone warning. This study corroborates the findings of earlier research (Sharma et al., 2009; Sharma et al., 2013).

Table 4.3 Reasons for fish decline and reduction in fish catch: fishermen's perceptions

Both overfishing and climate change effects			*Only overfishing*			*Only climate change effects*			*Denial of climate change effects*		
SF	MF	YF	SF	MF	YF	SF	MF	YF	SF	MF	YF
58	13	12	7	3	4	2	0	0	0	0	1

SF: Senior fishermen; MF: Middle aged fishermen; YF: Young fishermen.

The long-term impacts of the 2004 Indian Ocean Tsunami disaster

All of the fishermen respondents expressed the belief that the indirect impacts of the 2004 Indian Ocean Tsunami continue to influence and alter their fishing operations and fish populations in a variety of ways. They expressed concern about the declining fish populations and stocks over the past 15 years, especially since the 2004 tsunami. According to them, rampant capital-intensive fishing methods, the long-term effects of the 2004 Indian Ocean Tsunami, and climate change will force them to live in precarious situations in the coming years. Most senior fishermen agreed that recent weather patterns had been more abrupt than in the previous three to four decades. Fishermen of all socioeconomic backgrounds use the 2004 Indian Ocean Tsunami as a "time-yardstick" to compare changes in their livelihoods and fishing activities over the years. Multiple indirect effects of the tsunami still haunt their livelihoods. A fisherman about 65 years old talked about how fishing has changed in the last 15 years.

> Cyclones and storms weren't new, but the tsunami was. The tsunami shook "normal" weather and fishing seasons and changed the sea so much. Since the tsunami, commercial fish have moved to the deep sea/other coasts. Over the last 20 years, fishing has become more uncertain.

Although arguments linking the 2004 Tsunami disaster to climate change are wholly scientific and outside the focus of this study, fisher respondents constantly brought up the disaster's long-term ramifications, recounting how the tsunami severely impacted their fish harvest, coastal habitat, and radically changed weather and temperature patterns. According to fishermen's accounts, "The 2004 tsunami fundamentally transformed the coastal ecosystem, and there is no going back." Scientists are increasingly stressing and agreeing that rising sea levels due to global warming cause natural disasters such as earthquakes, submarine landslides, tsunamis, and volcanic eruptions (McGuire et al., 2002; McGuire, 2010, 2012; McGuire & Maslin, 2013). The intriguing connections between the rising and fluctuating sea levels and the disintegration of volcanic ocean islands with the rising intensity and frequency of tsunami disasters were discussed by McGuire et al. (2002, pp. 112–132). McGuire's (2012) work *Waking the Giant – How a Changing Climate Triggers Earthquakes, Tsunamis, and Volcanoes* details how global warming contributes to, triggers, and intensifies the complex interactions and effects of earthquakes, tsunamis, and volcanic eruptions around the world. In a similar vein, Li et al. (2018) demonstrate that even a modest sea-level rise has the potential to trigger the occurrence of tsunamis and could shorten the return period of potential future tsunamis, which may pose inundation threats to coastal communities. On the other hand, Spencer (2008, pp. 17–18) voiced strong opposition to linking climate change and tsunamis. However, there is no such comprehensive account of scientific studies focused on the Bay of Bengal's coasts that discusses the relationship between climate change and the frequency and severity of tsunami disasters. Therefore, in the context of the Bay of Bengal, there is a need for additional scientific studies examining the relationship between climate change and the occurrence of coastal hazards such as tsunamis.

Fishermen on climate change: perceptions through field narratives

Because of their familiarity with the sea and long history of working the ocean, fishermen have a unique understanding of the effects of various climate factors. Overfishing was reported as the most pressing threat to the livelihoods of many younger and middle-aged fishermen. In contrast, all senior fishermen had been aware of the varying environmental conditions and their negative impacts on livelihoods for at least three to four decades. As noted in the early portions of this chapter, it agrees with Santha (2008), who found that fishermen's local knowledge varies by their gear, their families, and where they live. In addition, this book also demonstrates that the age groups of the fishermen are a significant factor in the way that they interpret the weather. Young and middle-aged fishermen's views on climate change varied. During the fieldwork, it was often observed that only long-time fishermen with more than 30 years of experience could describe how climate change affects the marine fishing sector. Young and middle-aged fishermen saw heavy rain and drought as climate change signs. Most of the time, they ignored wind and current shifts. Extreme weather events (mostly high and strong tides), erratic winds, a longer summer season due to higher-than-normal temperatures, drought, coastal erosion, and rising groundwater salinization have all been felt by the fishing community due to climate change. During focus group discussions, senior fishermen often pointed out the multiple effects of climate change on increasing their risks while fishing and altering their normal fish catch patterns. All senior fishermen who participated in focus group discussions were aware of the changing environmental conditions over the past three to four decades and their negative effects on fishery livelihoods and coastal spaces. Seasonal anomalies and wild weather patterns have reduced fishing days and increased risks in the last 15 years. Turbulent and rough seasons have reduced their fishing days in the past decade. Using field observations, Figure 4.1 attempts to depict senior fishermen's climate perceptions.

Depletion of fish stock and less fish catch over the years: fishermen's experiences through the field narratives

The few climate change effects cited by fishermen were unpredictable weather patterns (mostly high and strong tides), irregular wind patterns, erratic monsoons, extreme summer temperatures (and the extension of the summer season), drought, coastal erosion, and rising groundwater salinization. Fishermen's perceptions of sea-surface temperature and its effects on fish reproduction and migration largely corroborated scientific findings (Kizhakudan et al., 2014). Pattinavar fishermen rely on Indian oil sardines

74 *Indigenous knowledge systems in confronting climate change*

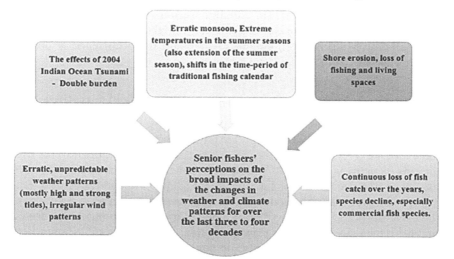

Figure 4.1 Senior fishermen's perceptions of climate change.
Source: Picture credit: Devendraraj Madhanagopal.

year-round (*Sardinella longiceps*). Prior to the 1990s, Indian oil sardines (Tamil: *mathi*) were unknown to the Tamil Nadu coast. Since then, oil sardines and other sardines have dominated the Tamil Nadu coast's fish catch (Murugan & Durgekar, 2008; Kizhakudan et al., 2014).

Fishermen's perspectives on fish stock depletion were evenly distributed across the survey sites, with all agreeing that fish stock in the sea and their catch levels have plummeted over the last two decades. Senior fishermen noted they can't find half the species they used to catch in traditional marine fishing grounds 30–40 years ago. They reported that Lady Fish/Indian Whiting (*Sillago indica*), Indian Anchovy (*Stolephorus indicus*), and Silver Moony/Silver batfish (*Monodactylu sargenteus*) are extinct or near extinction. Their primary source of income of marine fishers across the entire Coromandel coast of Tamil Nadu is the capture of Indian oil sardines (*S. longiceps*), Indian mackerel (*Rastrelliger kanagurta*), Seer fish (*Scomberomorus guttatus*), Prawn (*Fenneropenaeus indicus*), Little Tunny (*Euthynnus affinis*), and flying fish (*Cypselurus coromandelensis Hornell*). Senior Pattinavar fishermen said seasonal anomalies and wild weather patterns have reduced fishing days and increased risks over the past 15 years. In addition, rough weather has reduced their fishing days in the past decade. Some elderly fishermen also correlated the rise in temperature with the decline in fish catches, citing the increase in hot days as a primary cause. Several authors (Vivekanandan & Jeyabaskaran, 2010 (Indian coasts); Vivekanandan, 2011a, 2011b (South Asian and Indian coasts); Kizhakudan et al., 2014 (Tamil Nadu coast); Zacharia et al., 2016 studied climate change's effects on marine resources and fish productivity along Indian coasts. These works help to comprehend how climate change threatens Indian marine fisheries.

When asked about the fish catch and marine resources, a Kodiyampalayam fisherman said:

> In the past two decades, the depletion and decline of the fish population have been tremendous. If we catch enough fish to feed our family, we consider the fishing day fortunate.

Similar narratives were also shared by the Koozhayar fishermen. In this vein, the secretary of the Chinnamedu fishermen's cooperative provided additional information about the depletion of marine resources.

> In the past two to three decades, the decline of certain fish species has accelerated. If we catch these fish near the seashore, we consider the fishing day "very lucky". Currently, sardines, flying fish, Mackerel, and crab provide the majority of our income. During the rough season, we use our catamarans to fish in the nearby Kollidam River. There is no certainty in our income.

Surprisingly, the surukku valai fishermen of Kottaimedu and Madavamedu also bemoaned the decline in fish catches over the past few years and the fact that their catch does not cover their expenses, causing them to incur debts with moneylenders. When asked about the decrease in fish catch, a surukku valai fisherman from Madavamedu said the following:

> The lives of surukku valai fishermen are not as easy as others believe (he indicated small-scale fishermen of nearby fishing villages, including Kodiyampalayam). We must cover numerous expenses. Due to the ongoing depletion of marine resources by trawler fishermen, we, the surukku valai fishermen, have not been able to catch and sell enough commercial fish species in recent years. Consequently, there are several obstacles we must overcome in order to keep our costs under control. Due to the high accumulation of debts and less fish catch, many surukku valai fishermen from our village have already abandoned this profession and are now working as migrant laborers abroad.

A key informant in Kodiyampalayam explained the lower fish catch and species decline since 2004. For over 35 years, he has been an active fisherman.

> The decline of fish (especially commercial fish) has been severe in the last two decades. Before 20 years, we could catch commercial fish like Seer fish, Kingfish, and Pomfret near the shore. In the past, we regularly caught ray fish, little tunny, Pomfret, Indo-Pacific king mackerel, sardines, and seabass. Now, everything is very different. We can't catch much commercial fish even after traveling 20 or 30 km. On an average day, I fish once. Small-scale fishermen leave before dawn and return by noon. After that, we had to repair the fishing nets for the next day. We

cannot catch as many fish as we did 10 or 20 years ago, so we fish twice daily. Despite two daily fishing trips, we cannot catch enough fish to feed our family. We sometimes return with no fish to feed our families. Matthi (Indian oil sardines) is the only fish we can catch. More importantly, we are unsure how much longer we will be able to catch "Matthi" due to the constant "cleaning" of its stock in the sea by surukkuvalai and trawler fishermen. In addition, the quantity of fish we caught ten years ago has dropped dramatically. During fish-ban and rough seasons, we fish in the Kollidam River. Due to declining fish catches, we have started fishing during normal seasons.

Fishermen also pointed out the names of a few fish species, lamenting their extinction over the past five years. As there is no assurance of a fish catch, fishermen who operate outboard-powered vessels are unwilling to incur labor costs. In addition, they divulged that they have begun to spend time, fuel, and unnecessary effort searching for probable fish shoals in the ocean, even though a good fish catch is not guaranteed. Some of them recounted their encounters with huge, powerful gusts and whirlwinds while out on the water, highlighting the dangers of employing small boats for fishing in the inner sea. In recent years, fishermen have been unwilling to invest additional funds in fishing. Unless they have no other choice than to do so due to pressing financial needs, they have no intention of traveling into deep sea waters. All the fishermen in the villages under study mostly engage in near-shore fishing (who go fishing within around 10–20 km of the seashore). To catch "flying fish" ("*kola meen*" in Tamil), fishermen use small FRP boats and catamarans to venture into the deep sea (50–70 km) between May and August.[4]

Particularly, key informants feared that the precarious position will deteriorate in the near future. Rajshekar (2016)[5] examined the severe fall of fish capture along the coast of Tamil Nadu, citing multiple causes, such as climate change and overfishing. However, conversations with small-scale fishermen and observations in relatively large-scale fishing villages refute Rajshekar's claim. Rajshekar claims that large-boat fishermen are equally vulnerable as small-boat fishermen. Field observations and interviews with the fishermen of Kottaimedu and Madavamedu reveal that "large" fishermen (including surukku valai fishermen) are relatively less vulnerable than small-scale fishermen (fishermen who use FRP-OBM units and FRP catamarans) as they possess a limited stock of social and economic capitals to withstand financial turbulence. In Chinnamedu, the small-scale fishermen who fish from FRP catamaran boats and catamaran boats (without engines) reflected Rajshekar's results (2016). As previously pointed out, the small fishermen who fish in engineless FRP boats strongly perceive that weather and climatic patterns have become unpredictable and erratic over the past decade. However, they shared diverse responses to the reasons for the decline of marine resources, with climate and weather patterns being one of the fundamental causes. Throughout the research region, it was frequently observed that surukku valai fishermen were reticent to express their views on overfishing, as surukku valai fishing has been criticized[6] by the small-scale fishermen of the surrounding villages. The fishermen of Kottaimedu and Madavamedu, who are involved in surukku valai fishing, exhibited such reluctance (purse-seine fishing). Reduction of fish catch, species decline, and erratic weather patterns were not the only climate perceptions of fishermen. Climate change gives multiple and complex challenges to the fishermen. For example, a surukku valai fisherman of Kottaimedu linked climate change and coastal erosion with the loss of coastal spaces.

> During periods of intense precipitation and storm surges, boarding the boats has been extremely challenging. Due to climate change and the resulting coastal erosion, we are losing coastal areas and it is becoming more difficult to board boats. We can't board boats on the concrete because it will damage them. Our boats have been damaged while boarding during storm surges, causing heavy debts and a livelihood crisis.

The erosion of coastal spaces over time has a multiplicity of effects on the vulnerability of marine fishermen. Madhanagopal (2018) explains how the deterioration of coastal areas due to erosion exacerbates the life-threatening situations of Chinnamedu fishermen during climate extremes, since they knowingly accept risks to remain in fishing villages despite weather warnings in order to safeguard their fishing vessels and gears. Not only does the shrinking amount of space along the coast force fishermen to do less fishing there, but it also puts them at risk often.

Figure 4.2 The fishing vessels and shrinking coastal spaces. Place: Kottaimedu coast.

Source: Picture credit: Devendraraj Madhanagopal.

According to senior fishermen, the northeast monsoon has become more unpredictable in the past five years. Senior fishermen said the northeast monsoon is being delayed or stretched out. In Tamil Nadu's coastal districts, the northeast monsoon starts in October and ends in December. According to them, heavy rains from the northeast monsoon (November–December) are a precursor to a great fish catch. In the last 10–15 years, fishermen caught fewer fish in March and April due to a shift in the northeast monsoon's period. Few respondents in 2015 cited the delay and prolonged northeast monsoon as a reason for postponing peak fishing season to April and May when trawlers and mechanized fiber boats are banned. For roughly ten years, fishermen said "*vaadai*" is more unpredictable. Notably, most experienced fishermen observed that the onsets of "*vaadai*" have become more abrupt than their "normal" timeframe. Senior fishermen attributed the change in "*vaadai*" periods to the preceding decade, noting that as a result, they lost fish catches in March and April. A few of them cited the time-period shift of the northeast monsoon in the previous year (2015) as the primary reason for delaying the peak fish-catch season in April and May, which is the ban period.

Even during the fishing bans, small fishermen across the study sites continue to fish.[7] Strange weather patterns and the changing of the seasons have altered their fishing over the years.

According to one senior fisherman of Madavamedu, the current situation is as follows:

> Due to seasonal anomalies and erratic shifts in monsoon periods, our fishing days have been drastically reduced in February and March over the past five years. The normal fishing season for fish has been postponed until April and May, which is the fish ban period.

Intriguingly, a few fishermen suggested that, rather than implementing a 61-day fish ban at a time, the government should implement the annual ban twice a year. However, these viewpoints were negligibly rare. It should be noted that artisanal fishermen continue to engage in fishing with their small vessels during the ban period. However, they reported that it is useless to fish in the sea within 2–3 km of the coast because they cannot catch enough fish to sustain their families. A small-scale fisherman of Kodiyampalayam said:

> I fish even during the fish-ban months of April and May because I only use a small FRP catamaran. However, small-scale fishermen are allowed only to do near-shore fishing (1 to 2 km). We obviously can't catch any commercial fish species in near-shore fishing. Consequently, I typically fish the nearby Kollidam River during the fishing ban. In general, we get enough fish caught in February and March. However, for the past four or five years, our fishing activities in February and March have been severely disrupted due to changes in the monsoons and "*vaadaikaathu*" Normal fishing catch seasons have been postponed to April and May. However, we cannot use our motorized boats during this time due to the ban!

Approximately 40 fishermen who fish from small boats, such as FRP catamarans and catamaran boats (without engines), shared the same viewpoints. Both interviews and a focus group discussion confirmed this. During the summer, they typically catch crab. Due to the increasing complexity of weather patterns over time, the fish ban is currently causing difficulties. They reported that they could not catch as many crabs as they had in the past five to ten years. It is summer in India from March to June. All senior fishermen observed that maximum summer temperatures were lower around 10–15 years ago than they are now. They pointed out that the severity of heat during June and July for the past five years is nearly identical to "*Agni nakshatram*": In Tamil (the peak summer period that happens in May). Over 105 years (1906–2010), a rise in sea surface temperature along the coast of Tamil Nadu had a direct correlation with a decline in fish catches (Kizhakudan et al., 2014). Fishermen believed that short-term changes in weather patterns and decadal variations substantially impacted marine resources. They always mentioned the changing seasons and the rise in temperature over the past five years, and they explained how these factors affected where the fish lived, where they reproduced, and how many fish there were. A catamaran fisherman from Chinnakottaimedu with more than three decades of fishing experience described how the erratic rise in temperature over the past three decades had impacted fish catching.

> The temperature must be conducive to fish shoals in order for us to catch enough fish. If the sea is warmer, fish shoals travel to deep-sea or some other fishing grounds where the sea is less warm. Due to this, we decided to travel more frequently by sea, which is extremely dangerous.

A few senior and younger fishermen across the study sites believed that weather and climate variations over the years were merely cyclical phenomena and that nothing "new" had occurred to them. A few of them mentioned that large-scale fishing had been a major contributor to weather and climate changes in their region over the past few decades. It concurs with recent research conducted in the Peruvian Andes (Paerregaard, 2018). Over the course of more than three years (May 2015 to May 2017), systematic follow-ups of climate observations and fishermen's perceptions in the study region reveal that fishermen's perceptions of their local environment changed in tandem with weather patterns. In 2017, fishermen's distress over changing weather patterns was more complicated than in the previous two years (2015) and (2016). In 2017, the same fishing villages were revisited, and a significant number of respondents were interviewed to determine the consistency and variances in their perceptions of climate change. In November, December 2015, and January 2016, throughout the study area, fishermen experienced heavy rainfall. These events could also shape and influence their perceptions of weather and climate stress. This finding partially backs up Paerregaard's research (2016). Paerregaard found that Andean community members in Peru recently begun to perceive the local effects of climate change in various

Table 4.4 Comparison of climate perceptions of fishermen with the recent scientific findings[8]

Climate variables	Climate perceptions of fishermen	Recent scientific findings	Conformity
Sea-level rise	Fishermen possess limited knowledge to discuss sea-level rise.	Five coastal districts of Tamil Nadu, including Nagapattinam district, are highly vulnerable to sea-level rise, and they are estimated to be the likely impacts of storm surges (Byravan et al., 2010).	Unknown
Shore erosion	Heavy for over the past three decades. It has been very heavy since the 2004 Indian Ocean Tsunami disaster.	Tamil Nadu's coasts are eroding. In addition to climate change, rapid urbanization and industrialization contribute to rising shore erosion (Natesan et al., 2015; Bhalla et al., 2008).	Agreement
Temperature	Increasing trend. These increasing trends have direct impacts on fish stock and	Since 1901, Tamil Nadu has had more dry days than wet days each year (Guhathakurta et al., 2011). Another recent study (Kizhakudan et al., 2014) on sea surface temperature shows a rise in SST along the Tamil Nadu coast over 105 years (1906–2010). This is evident across the coastal regions of Nagapattinam district for over the last 20 years (1990–2010).	Agreement
Rainfall	Decreasing trend and increasingly uncertain.	From 1970 to 2000, Tamil Nadu saw heavy northeast monsoon rains and a slight decrease in southwest monsoon rains (Kumar Bal et al., 2016). From 1970 to 2000, rainfall projections demonstrate a declining trend (CCC&AR & TNSCCC, 2015). Northeast monsoon, the primary rainy season (October–December), is increasingly chaotic (Indira & Stephen Rajkumar Inbanathan, 2013).	Partial agreement

(*Continued*)

Table 4.4 (Continued)

Climate variables	Climate perceptions of fishermen	Recent scientific findings	Conformity
Climate events	Increasing trend. It's becoming more uncertain over the past three to four decades. However, fishermen were not able to point out the exact changes in trends according to the seasons.	Between 1891 and 2006, the coastlines of Tamil Nadu were affected by around 32 cyclones, 30 of which were severe. By 2011, the total number of cyclones that impacted the Tamil Nadu coast had risen to 44. As a result, between 2006 and 2011, there was a 37.5% rise. Increasing trends of cyclonic disturbances in the Bay of Bengal were observed during the winter (October–February) and pre-monsoon months (March to May) (Government of Tamil Nadu, 2013).	Partial agreement

ways, based on his 30-year (1986–2016) study with these community members. According to Paerregaard (2016), people in the Andean community believed that climate change and the resulting temperature rise were caused by their modern lifestyles and the pollution from their local industries, not by an irreversible global phenomenon. The small sample of marine fishers in this study shared these concerns. However, they numbered far fewer people than those who held the more widespread belief that climate change would have severe negative consequences for their fishing livelihoods. Even though the follow-ups that this book presents based on the study occurred at much shorter intervals than Paerregaard's, this analysis highlights similar reflections in the broadest sense, namely, fishermen have begun experiencing the localized impacts of weather and climate stress over the years.

Potential life-threatening situations due to erratic weather patterns

Fishermen categorize sea currents according to the seasons by local dialects. They have a substantial body of knowledge about "*vaada neerottam*" (sea currents that occur in rough seasons). It gets strong in November and December. Fishing in November and December using non-motorized vessels is a risky job. To navigate the vessels safely, fishers should have skills from traditional wisdom to construe the nature and speed of sea currents. Hence, fishermen are less likely to venture out to sea during stormy weather. Even if they go fishing, they prefer fishing close to shore. The findings indicate that fishermen strongly believe in erratic changes in the patterns of winds and sea currents;

consequently, a variety of life-threatening risky situations have made fishing occupations increasingly precarious. According to them, wind patterns and ocean currents are the most influential determining factors on any fishing day (Fieldwork; Samas, 2015). During focus group discussions, Kodiyampalayam and Chavadikuppam fishermen recounted numerous personal encounters with erratic winds and ocean currents. Since they fish in both FRP-OBM units and engineless FRP boats, they have had extensive direct experience with strong waves and erratic sea currents. As described by a young fisherman in Chavadikuppam village, who is approximately 40 years old:

> Over the past twenty years, I have been an active fisherman. Fish species are migrating to the inland sea and other parts of the sea as a result of the lengthening summer season and the increasing number of hot days. Over the past five to ten years, we have experienced *"vaadai kaathu"* in January and February because of increasing seasonal shifts; *"vaadai kaathu"* is an indicator of rough sea conditions. It is risky to fish in the *"vaadai kaathu"*. Due to this *"vaadai kaathu"*, I found myself in a perilous situation at sea in February of 2016. My boat capsized after being struck by a powerful wave. In the water, I drowned. I was able to make it to the shore, but I lost my boat and nets. As far as I am aware, the government does not offer compensation for vessels and nets damaged by severe weather. In a few instances, they (government and NGO representatives) say something about insurance programs and all… However, I am not aware of this. No one has come to us for help obtaining a government compensation amount, as far as I am aware. At the end of the day, we, fishermen take the loss, and these occurrences occur frequently in our lives.

Excessive rains for more than three months (November–December 2015 & January 2016) severely affected the fishing activity of fishers across coastal Tamil Nadu for around three months. The fishing vessels and nets of fishermen were damaged by the severe rainfall. Many elderly fishermen across the research region remarked that the frequency and intensity of the rain had been lower in the last 15 years than in prior decades. In small fishing villages like Chinnakottaimedu and Chavadikuppam, these voices are louder and deeper. However, during the pilot work, it was found that observations of fishermen on rainfall and climate events were frequently discordant with one another. They have a limited capacity to recall the specifics of climate events and rainfall intensity during the past three to four decades. Table 4.3 attempts to compare the climate perceptions of fishermen with the recent scientific findings. It shows that the climate perceptions of senior fishermen largely concurred with the recent scientific findings.

Weather updates and early warning systems
Over the past ten years, fishermen have increasingly becoming more reliant on cooperatives for early warning information systems. Regardless of socio-economic and political variations within the research area, fishermen value

the fishermen cooperative society as the major and authentic source for weather updates and early warning information. In addition, the M.S. Swaminathan Research Foundation (MSSRF) and the Reliance Foundation provide frequent weather-related updates to the fishermen. Such external sources supply regular weather updates to the registered cell phone numbers of fishermen, which they find valuable for their daily fishing activity. It should be noted that external assistance is becoming increasingly scarce and fragmented. For instance, fishermen in Chinnamedu reported receiving weather updates and early warning information from MSSRF, whose district head office is in Poompuhar, Nagapattinam district, Tamil Nadu, for the previous three months. They all agreed that this outside assistance had been extremely beneficial to them in both their daily fishing efforts and during the rough seasons. Such external weather-related information support for the district's fishermen is, however, inconsistent, and fragmented. The fisheries of Kodiyampalayam and Koozhayar lack these weather-related external information support systems. Hence, as this book shows, small-scale fishermen who reside in villages along the Coromandel coast without external weather-related support systems rely heavily on their local (traditional) body of knowledge.

Indigenous knowledge in the context of climate change

The majority of fishermen (almost 90%) rely increasingly on the media for early warning information, especially after the tsunami. Aside from this, a few private actors in the research setting have been very important in giving fishing villages regular updates on weather information. However, it is not uniformly distributed. They provide weather updates via text messages and phone calls to the registered mobile phone numbers of fishermen. It has been quite beneficial, according to the fishermen. Due to technological advancements and the increased use of mass media among fishermen, especially in the wake of the 2004 tsunami, the likelihood of losing lives due to climate catastrophes has been nearly eliminated. However, because of climate change, fishermen regularly encounter a variety of obstacles and threats. Small-scale fishermen confront two types of issues as a result of climate change. That is, while fishing, they face (i) trouble locating and assessing potential fishing zones in the ocean and (ii) difficulties in safely navigating their fishing vessels under extreme weather conditions.

According to the secretary of the Kodiyampalayam fishermen's cooperative group,

> In the past 30 years, we have been subjected to a variety of climate pressures, and their frequency is increasing. We can no longer rely solely on local ecological knowledge to predict environmental precursors, fish distribution, and abundance since we have repeatedly realized and experienced that our traditional knowledge is no longer precise and sufficient to understand coastal systems.

When questioned about how unpredictable big waves could be an added strain to their lives, a senior Chavadikuppam fisherman responded:

> Our local traditional expertise for detecting weather factors such as wind speed and direction, sea current speed and direction, water mass movement, and upwelling is gradually becoming obsolete, hence compounding our livelihood crisis. Significant commercial fish species have already migrated to the interior sea. As a result, we are compelled to drive to the inner sea to capture commercial fish species, which will demand additional fuel, labor assistance, and will be more dangerous.

Similar perceptions were shared by the middle-aged and young fishermen across the study sites of fishing in the interior sea by small fishing vessels. It suggests that small-scale fishermen risk their lives to earn their daily living. In many instances, fishermen have no idea how to respond to the problems posed by erratic weather patterns and accompanying climate and weather impacts. They realized that their prior experiences with the sea and weather patterns were of moderate value under the changing environmental conditions. Prior to roughly two to three decades ago, fishermen never felt the need for "science" in their daily fishing endeavors. However, over the past two decades, they have begun to acknowledge the need for scientific updates and weather and navigation instruments, especially after the tsunami. Thus, fishermen's views tend to rely more on scientific approaches than merely on their traditional knowledge. Most fishermen who participated in the survey, regardless of age, education level, or ownership of boats/nets, anticipate assistance from external sources (including state administration and donor agencies) to forecast environmental variations and the movement and abundance of fish shoals. The growing frequency of variable weather patterns has rendered traditional seasonal calendars less reliable for fishermen. In addition, as this book shows, due to technological advancements, fishermen no longer rely on their traditional knowledge to identify environmental precursors. It is congruent with the findings of Nayak (2017), who notes that young fishermen in Odisha had lost interest in acquiring traditional knowledge and skills. As a result, there is a need to systematize and protect small-scale fishermen's indigenous knowledge. Kurien (1996) writes that artisanal fishermen's knowledge, built over centuries, will be re-energized and made stronger by scientific research. Participants in the study shared the same observations, stating that they believe both short-term changes in weather patterns and decadal variations have direct and indirect effects on marine resource. They blamed the shifts in weather patterns, specifically the changing seasons and the rise in temperature during the previous five years, and explained how these factors had influenced the reproduction and breeding patterns of fish populations. A catamaran fisherman in Chinnakottaimedu recounted how temperature increases affect fish catches. He was 45, and he had over 25 years of fishing experience.

If the temperature is favorable for fish shoals, we have a good fish catch. If the sea is warmer, fish shoals will migrate to the deep sea or another shore where the sea is less warm. My small boat cannot venture more than four or five kilometers out to sea to gather fish, as it is too dangerous.

From the study, it is clear that most young and middle-aged fishermen, in contrast to their forebears, look to modern scientific methods to guide their fishing rather than relying solely on ancient knowledge. Overall, fishers, regardless of age, literacy, or boat/net ownership, largely look forward to external support to understand weather and climatic patterns and also to face adverse climatic conditions as they believe their local environment has already changed (Madhanagopal & Pattanaik, 2020). Hence, by pointing out the insights and observations of this chapter, this book puts forth a claim that the fishers' traditional knowledge of oceanographic factors and marine ecosystems may not entirely suffice to confront the existing and upcoming climate change challenges. Therefore, as Wisner (2009) argues that there is a ground necessity to consider validating and embedding their local knowledge in a scientific context; rational assessment must be done to substantiate its effectiveness in the present-day context. The recent studies discussed the need for integrating local knowledge for climate change adaptation and disaster risk reduction (Raymond et al., 2010; Kelman et al., 2009; Armitage et al., 2011). However, integrating local and scientific knowledge is a long process, and many factors and contexts have to be examined (Williams & Hardison, 2013; Hiwasaki et al., 2014). Alexander et al. (2011) suggest that traditional knowledge and indigenous narratives help to a sufficient understanding of global warming. Incorporating such indigenous narratives with scientific knowledge represents a viable method for assessing the effects of climate change and recognizing climate adaptation efforts.

Notes

1 A substantial portion of this chapter was published in the journal *Environment, Development, and Sustainability* under the title "Exploring fishermen's local knowledge and perceptions in the face of climate change: the case of coastal Tamil Nadu, India." I am thankful to the editors of this journal for granting me permission to reuse the published article. I also want to point out that this journal article only focuses on three fishing villages. On the other hand, this chapter focuses on all seven fishing villages and expands on the discussions on indigenous knowledge systems and fishermen's perceptions of climate change. Overall, the fishing experience and age of the fishers' respondents that this book focuses on slightly differs from this published article. The published journal article mainly focused on senior fishermen with 30–45 years of fishing experience. However, as the discussions of this chapter mainly center on indigenous knowledge systems and climate change perceptions of the senior Pattinavar fishermen, there is no substantial difference arising from this chapter and the published article.
2 The "community lingo" of fishermen is transliterated from Tamil to English and italicized.

3 Fishers of the coastal districts of Tamil Nadu consider October to November the "rough season." The Northeast monsoon starts in October and ends in November, contributing 60% of the total annual rainfall to the state. Nagapattinam district is frequently affected due to the Northeast monsoon, which leads to cyclonic storms, flooding, and massive rains at least once every three or four years.
4 The fishermen's fish catch and the month details were not consistent. The fishermen roughly described the months in which they capture "*Kola meen.*" In addition, fishermen discovered a correlation between the weather patterns and the catch of fish (*Kola meen*) and reported that the patterns had become increasingly unpredictable in recent years.
5 This article discusses the ongoing fall in fish catches along the coast of Tamil Nadu since the 2004 Indian Ocean Tsunami, as well as the vulnerability of small fisherfolk to this trend. Increasing debts, dwindling fish catches, and escalating expenses entrap traditional fishermen and force them to leave fishing for the most part. Climate change, overfishing, the use of prohibited trawling gear, pollution, and intensive fishing make the livelihoods of small fishermen even more precarious.
6 Kodiyampalayam, Chinnamedu, and Chavadikuppam small-scale fishermen stated that surukku valai fishing is as destructive as trawler fishing since it makes it hard for them to make a living.
7 The period of the fishing ban differs between India's west and east coastlines. This seasonal closure usually commences from April 15th to May 29th every year (though it has been extended until June 14th). This ban applies to all mechanized fishing vessels, including purse and ring seines. In the field visits, it was found that few fishers use their OBM boats to do near-shore fishing. During the ban, traditional, non-motorized vessels are allowed to fish along the coast. The state government has increased the relief assistance for fishermen (during fishing ban periods) from INR 2000 to INR 5000 beginning in 2017–2018. In addition, starting with the 2016–2017 fiscal year, the state government has enhanced the special allowance scheme for marine fishing families from INR 4000 to INR 5000. But fieldwork in 2017 shows that fishermen are still unhappy with this new assistance amount because they do not think it is enough to support their families during this period.
8 Similar comparisons were done and published in the journal *Environment, Development, and Sustainability*. Only senior fishermen (n = 40) were compared to the scientific findings in the published paper. In contrast, this chapter draws parallels between how the fishermen in all the study villages perceive climate change and the scientific findings. However, there is no discernible difference between the results obtained.

5 Living with climate change
Vulnerability and adaptation actions

Social vulnerability to climate change

Vulnerability is temporal and spatial and depends on social, economic, cultural, geographical, and political factors (Cutter & Finch, 2008; Cardona et al., 2012). Individual and social factors determine climate change vulnerability. Income levels, asset holdings, education status, social class, gender, age, livelihood circumstances, and ethnicity are some elements that impact vulnerability and adaptive capacity to climate change. Some social groupings have distinct coping techniques, and their reactions vary by nature and power. Adaptation to the effects of climate change is nothing new, and it has naturally existed among communities in diverse ways (Adger et al., 2005; Mearns & Norton, 2010). Fishing communities in the developing world face various challenges, including poverty and restricted access to resources that make them more vulnerable to climate change and extremes. Vulnerability to climate change and the role of institutions in fostering adaptive capacity can only be better understood by first examining how vulnerable local people are to it. Ribot (2011) stresses that understanding the origins of vulnerability can help us identify the social, economic, and political actions required to address the challenges of the vulnerability of the victims. Hence, vulnerability analysis is the first step for adaptation interventions. Fishers' responses to climate change differ depending on their potential for adaptation. Also, as the previous studies on fisheries point out, the asset base of the fishers, including their social, economic, political, and institutional capital, decreases their vulnerability to climate change (Badjeck et al., 2009; Islam, 2013).

Geographical isolation

The geographical isolation of fishing villages and the limited accessibility of public transport facilities make Pattinavar fishers more vulnerable to climate change. In particular, Chavadikuppam, Chinnamedu, Chinnakottaimedu, and Kottaimedu fishers have limited access to public transportation. The distance between the main roads to Chinnakottaimedu and Chinnamedu is around 4 km, making poor fishers, especially fisherwomen, more affected

DOI: 10.4324/9781003187691-5

during extreme climate conditions. The geographical location of Kodiyampalayam makes it almost appear like an island. Its geographic location and remote nature make the fishers more prone to the impacts of climate change and extremes. However, this village has adequate public and private transportation facilities compared to the nearby small fishing villages. As field observations show, there are no separate cyclone shelters for Chavadikuppam, Chinnamedu, Chinnakottaimedu, and Kottaimedu. Still, there are shelters in the nearby villages, which are around 3–5 km away from these villages. Hence, the villagers of those villages must travel a few km to reach cyclone shelters during climate events, which is riskier. Hence, in many cases, the fishers do not prefer to travel and stay in cyclone shelters; they prefer to stay at their own houses. Moreover, cyclone shelters are too small to accommodate the whole village. In the case of Chinnamedu, there is a cyclone shelter in the nearby village. The distance to reach the shelter from the main village is around 5 km. It should be noted that this shelter is a place that is expected to be used for three villages; Chinnamedu, Chinnangudi, and a nearby village, which is not a fishing village. The population of these three villages is approximately 4000. In contrast, the shelter can only accommodate a maximum of 400–500 people. Aware of these space constraints, Chinnamedu fishers normally avoid residing in cyclone shelters in their villages during cyclone periods and prefer staying in their homes. Such taking such risks are common among the fishers of Chinnakottaimedu and Chavadikuppam. A cyclone shelter is available in Kodiyampalayam, an old building that is too small to accommodate the whole village. However, these villagers have further complexities in shifting to cyclone shelters during erratic weather due to the geographical nature of this village. As the sea and the river on all four sides surround this village, the villagers of Kodiyampalayam continue to live in fear during the rough seasons. Around 50 fishing families in this village live approximately 300 meters away from the sea. During the rough seasons, the substantial fishing assets of these fishers are regularly damaged. On average, the fishermen who own fiber-reinforced plastic (FRP) catamarans and outboard motor (OBM) boats spend a minimum of Rs. 30,000[1] to repair their fishing assets. Besides, the fishermen have reported that their houses (including tsunami houses) are getting damaged during rough seasons. It is also noted that as fishers of these villages are more concerned with the damage to fishing assets, they give more priority to their fishing assets than their own houses. It indicates their social vulnerability to climate extremes. During the fieldwork, the respondents often noted that they expect support from the state and the NGOs to repair their houses since most of them are in free-tsunami houses. Chavadikuppam and Chinnamedu fishers reported many complaints about the poor quality of free-tsunami houses, noting that these houses are not sufficient to protect them in rough weather. All key informants of small fishing villages repeatedly registered their grievances about the poor quality of free-tsunami dwellings, which make them more vulnerable to climate extremes.

Sanitation and wastewater management

Fishing villages irrespective of their financial capitals lack effective sanitation and wastewater control measures. In particular, Chavadikuppam, Chinnamedu, and Chinnakottaimedu lack sanitation and septic systems, making them more susceptible to diseases during and after extreme weather conditions. Prior to the tsunami, they did not have separate toilets. Significant fishing households now have indoor toilets in their tsunami homes. However, because the indoor toilets in their little homes are uncomfortable, they typically avoid using them. Even during construction of tsunami homes, they attempted to persuade the contractors, but their recommendations were disregarded. Some of them had attempted to use the inside toilets for a few days, but they were all permanently closed. Senior fishermen are concerned about the deteriorating economic and infrastructure conditions in Nagapattinam's fishing villages. Duyne Barenstein (2016) demonstrates, through a case study of fishing villages in the district of Nagapattinam, how the state and non-governmental organizations violated the right of fishermen to live in adequate and decent housing. She raises the practical and ethical questions of sacrificing the immediate lives of relocated communities for the sake of possible future disaster events by highlighting the flaws in the state policy of putting more emphasis on building new houses for tsunami victims while not providing any compensation to build and repair their existing houses. Discussions with the fisher respondents on their housing and sanitation facilities echoed the discussions in the previous literature (Duyne Barenstein, 2009; Duyne Barenstein, 2016).

A lot has changed in this region since tsunami recovery funding started flowing through. However, in Chavadikuppam, Chinnakottaimedu, and Chinnamedu, majority of fishing households continue to practice "open defecation." There are several health problems that arise when fishers defecate in the open, especially when it rains heavily. Fisherwomen are particularly prone to such infections due to a lack of sanitation, particularly during stormy seasons. The torrential rains that fell in November and December 2015 had a significant negative impact on the fishers in Chinnamedu and Chavadikuppam. They had to rely on moneylenders for their financial demands because the stagnation of rainwater in villages disrupted their fishing activities for around ten days. Chinnamedu, Chavadikuppam, and Chinnakottaimedu are especially vulnerable as their households' wastes must be disposed of in their own backyards. They recognize the numerous hygiene problems it has brought about. But they have no other choices. Notably, during focus group discussions in Chinnamedu, fisherwomen prioritized health and sanitation issues more than fishing concerns. They also went on to note their weak financial and political capitals and pointed out that their problems are largely unheard by the state. Conversations with the fishers at different time intervals show that they typically anticipate assistance from external sources, either from the state or NGOs, to address such issues. The state and an NGO donated funding for the construction of private toilets for these fishers. Madavamedu and

Kodiyampalayam, which possess sufficient political and economic capital to make good use of such funding, fared well. However, as fisherwomen revealed, it did not perform successfully in Chavadikuppam and Chinnamedu due to the ignorance of local key actors. Interestingly, unlike the fishers of small-fishing villages, Madavamedu and Kottaimedu fishers are content with their free-tsunami homes. Many houses do not appear to be free-tsunami houses as they renovated, enlarged, and renewed their homes, including the entrance. Private toilet facilities in the houses of those fishers are still limited, indicating a lack of hygiene awareness. Nevertheless, fishers who do surukku valai fishing have toilet facilities in their homes. In general, the fishers believe that the attached toilets are uncomfortable and unusable to them. These two villages are financially stable compared to the others. Hence, it can be speculated that individual financial stability and collective self-organization are not reflected in the village's public sanitation. The fishers, largely, are prone to infectious diseases during and after climate extreme events.

Poverty and gender inequalities

This study supports Clark et al. (1998), who indicated that poverty and gender inequalities made the coastal population vulnerable to climate change and events. Poverty and gender discrimination are the major social vulnerability factors prevalent across the study region. In Chinnakottaimedu, Chinnamedu, and Kodiyampalayam, poverty among the fishers is evidently visible. Chapter 7 of this book details the ways in which gender inequality in the fishing villages and their governance systems impacts fishers' vulnerability and reduces the hold of women's agency. Also, as the fieldwork demonstrates, poverty and gender inequalities are not the only climate change vulnerability factors at play in the lives of Pattinavars. Many other factors corner the fishers to be more exposed to different ranges of vulnerabilities. For example, the vulnerability of the fishers is directly linked to the fish catch. Perhaps if the fishermen return from the sea with insufficient or no fish caught, the fisherwomen opt not to go to the adjacent villages to buy fish from them to sell and earn for their daily living. Instead, they decide not to sell fish on the streets that day. As a rule, they avoid buying fish from neighboring fishing villages because it can be complicated. These are the explanations that emerged from field insights: Not all fisherwomen must pay cash to purchase fish from fishermen who return from the sea. Mostly, marginal fisherwomen are involved in selling fish on the streets in nearby towns. Some families lost their breadwinners (typically spouses) due to the tsunami. Fisherwomen (head loaders) often range in age from 40–50. Every day, they make money and spend it. Due to their low income, they rarely put money aside for the next day. Hence, they cannot afford to pay cash for the fish in many circumstances. Instead, they pay a portion of the money as a debt to the fishermen and then refund it on the same day evening after selling the fish on the streets. Fisherwomen typically have mutually business understandings with the fishermen from whom they routinely purchase. Furthermore, the fishermen in Chinnakottaimedu and Chinnamedu have solid kinship and social ties, giving them access to a potential source of social networks. As a result, getting and paying back debt is now a part of their daily routine. Because of the lower or no fish caught on a certain day, they cannot change their transactions to the surrounding fishing villages. If they do, perhaps, it may cause social and commercial friction among fishers and fishing villages. Hence, fisherwomen avoid buying fish from neighboring fishing villages due to the cost. Instead of selling fish, they skipped work one day. Small fishing villages like Chinnakottaimedu, Chavadikuppam, and Chinnamedu have further compounded the situation. Their main source of income is small-scale fishing (no surukku valai fishing). They are all from small fishing villages surrounded by bigger ones, such as Paazhaiyaar, a village predominately engaged in capital-intensive surukku valai fishing. Thus, comparing the pointed study villages to the surrounding large fishing villages (Paazhaiyaar), one can see the dramatic discrepancies in infrastructure and road accessibility facilities amongst the fishing villages along Tamil Nadu (the distance is less than 5 km).

Fisherwomen typically lack their own private transportation service. They must either rely on the male members of their family or pay for specialized transportation, such as auto-rickshaws. In Chavadikuppam and Chinnamedu, fisherwomen who are involved in selling the fish rely on male family members for transportation. At the same time, the fisherwomen without male relatives (typically widows) face an extra burden. To reach the main road, they must walk. Because of their age, they normally avoid selling fish on the streets on any given day. Otherwise, to reach the main road, fisherwomen must walk 4 km

Figure 5.1 Fishing vessels of small-scale fishers. Place: Chinnamedu beach.
Source: Picture credit: Devendraraj Madhanagopal.

while toting more than five kilograms of fish baskets on their heads. Perhaps they could walk an additional 4–5 km to buy the fish from the adjacent villages if little or no fishing occurs on any given day. Thus, having little to no fish catch intensifies stress in the lives of fisherwomen. Therefore, the daily lives of fisherwomen are immediately impacted by species extinction and decreased fish catches. In particular, these incidents are brunt during extreme climatic conditions, and women without male family members are particularly prone in numerous respects. Access to facilities like electricity, drinking water, information, and communication is similar in the study regions. According to fishermen, the main difference in treatment is the compensation received during stormy seasons and cyclones – Only large fishing villages are privileged to have seawalls built on their beaches.

Other escalating issues that aggravate climate change vulnerability

Water salinity

Protests against the prawn industry have been ongoing for decades in the fishing villages of Nagapattinam district, particularly among the farmers directly affected by the tsunami. Senior respondents blamed that the shrimp industries along the Coromandel coast have already affected their livelihoods and increased the salinity of their water resources. In several cases, uur panchayats protested prawn and new shrimp and farm industries. In a plea to the collector, the uur panchayat leaders of Chinnamedu pointed out that the prawn industry was wreaking havoc on the local environment (Revathi, 2015). During the focus group discussions with the senior fishermen of Chavadikuppam and Chinnamedu on their beaches, they stressed that the shrimp and prawn businesses are to blame for rising water salinity over the previous two decades. Fishers across the Coromandel coast of South India rely heavily on groundwater for their daily operations, regardless of their socioeconomic status. A steady rise in salinity has been noted by fisher respondents in recent years. Except for those in Chavadikuppam and Chinnamedu, most fisherwomen had no idea why the water's salinity was rising. In the fieldwork, just five senior respondents made links between the increasing water salinity and climate change. Fishers in Chavadikuppam and Chinnamedu are aware of the rising salinity of their local drinking waters, which they attribute mostly to the growth of aquaculture ponds and shrimp farming in their regions.

However, winning against shrimp farming industries is challenging because those proprietors have direct ties to and support wealthy local individuals, politicians, and the local government. So, fishermen believe they cannot agitate and win successfully against them. Overall, the respondents cited three reasons for the rising salinity. They are as follows: (i) Shrimp farming, (ii) coastal erosion, and (iii) tsunami impacts. Fishermen were large of the opinion that the 2004 Indian Ocean tsunami disaster had adversely impacted their groundwater resources. The tsunami waves delivered a tremendous inflow of salty seawater into the fishing villages, which remained stagnant for days, damaging the hydrogeological systems (Chaudhary et al., 2006). A 70-year-old senior respondent in Chavadikuppam described the effects of coastal erosion on groundwater. He has been fishing in Chavadikuppam village for the last 45 years and is one of the direct witnesses to the 2004 Indian Ocean tsunami, shrimp aquaculture, and the effects of beach erosion on his village's livelihoods and groundwater. In addition, he served as the head of local institutions and has been an outspoken critic of shrimp farming. Not only a critique, but he also directly participated in a few agitations against shrimp farming some years ago. He has lived in the same village all his life and possesses adequate experience to determine the worst of shore erosion in his area. According to him, every natural disaster had long-term implications on their fishing grounds and groundwater resources.

> Only a handful of people my age live in my village. Over the past four to five decades, the beach spaces of our village have been steadily eroding. In the last 40 years, shore erosion has taken away about 30% of our fishing space on the back. Particularly, since the tsunami, the sea is inching towards the village faster. The sea's landward displacement over the past 12 years has been higher than in the last two decades. Seawater intrusion surged following the 2011 Thane cyclone and recent high rainfall (2015). Therefore, our groundwater is becoming more and more saline in recent years, resulting in uselessness. Most shrimp/prawn companies in the area, especially shrimp farming, are held by affluent people in other districts. This industry has been the subject of numerous protests. However, the local authority seems unwilling to listen to our complaints. They allowed shrimp farms to operate, damaging our ecology and livelihoods. Due to personal losses, most have ceased their industry within a few years.

In a similar vein, most respondents in small-fishing villages shared several complaints against shrimp farming. This is because small fishing villages such as Chinnamedu, Chinnakottaimedu, and Chavadikuppam are more vulnerable to such damage than their counterparts. Fishers of Kodiyampalayam, Madavamedu, and Kottaimedu are also equally vulnerable to climate change. They do, however, have specific economic and political capital, as well as social networks, to respond to climate stressors (Madhanagopal, 2018). Fishers belonging to small-fishing villages shared their grievances that the state administration and bureaucrats are often not ready to help them. They collectively noted that capital-intensive fishing (*surukku valai, rettai madi, iluvai madi*)[2] villages possess strong socioeconomic and political capital to negotiate with the state administration, and thereby, they receive benefits for their villages.

Complicated state climate events compensations

The government of Tamil Nadu continues to develop several welfare programs to assist marine fishermen with their lives and livelihoods. The fisheries

94 *Living with climate change*

policy note (2017 and 2018) states that during the fishing season, the fishermen's welfare board of Tamil Nadu typically provides 0.1 million INR to the families of missing or deceased fishermen (Government of Tamil Nadu, 2018).[3] In practice, however, the compensation amounts vary depending on the severity and current sociopolitical circumstances. For instance, the government of Tamil Nadu gave INR 2 million to the families of the fishermen who went missing or died during the Ockhi cyclone in 2017. Following the massive 2015 rains, many news outlets reported that the state administration made significant efforts to drain stagnant waters and assess the damage to marine fishers' assets. Nevertheless, a senior fisherman from Chinnamedu criticized state aid and shared contrasting perspectives.

> If any storm, flood, or heavy rainfall occurs, elected representatives and bureaucrats will visit our villages to evaluate the damage to assets; for us, it is a familiar story. But getting timely and fair compensation from the state is rare. It was the same story for the Thane cyclone – we did not receive due compensation. The sole exception was the tsunami disaster. Recovery and rehabilitation went very well in the tsunami disaster as many NGOs came and provided support in various ways.

Such perspectives align with other fishermen's accounts across the study region, especially among the fishers who own artisanal fishing vessels. Economically poor fishermen try to safeguard their assets from possible loss during climate events because they know that they cannot rely on the state's compensation. Widespread feelings of fear and insecurity have led them to

Figure 5.2 A small-scale fisherman is repairing his fishing net. Place: Chinnamedu beach.

Source: Picture credit: Devendraraj Madhanagopal.

Living with climate change 95

avoid emergency evacuation plans. During climate extremes, they (especially, poor fishing households) choose to risk their lives to protect their households and fishing assets (Madhanagopal, 2018). An uur panchayat leader of Chinnamedu explained how the fishers of small-fishing villages are more vulnerable to climate extremes.

> Our fishing population suffered greatly during the Thane cyclone in 2005 and the recent heavy rains (December 2015). Many of our homes and electronic devices were damaged. Even after the rain stopped, the rainwater stood in Chinnamedu for days. As is customary, the government officials delayed their visit to the village. We did not receive proper compensation from the state, which is no surprise. Every year during rough seasons, we are affected. Our villages lack net mending halls to keep nets safe. During the 2011 Thane cyclone, around 20 fishing households lost their assets. Coastal erosion has already shrunk our village's coasts, and we have fewer coastal spaces to protect our boats and nets.

Debt traps: lives that rotate the moneylenders

Fishers of small fishing villages across the study region resonated with such aforementioned views. A fisherman from Kodiyampalayam explained how the lives of fishers span around moneylenders and the ways in which climate change and associated risks affect their fishing activities. He owns an FRP-OBM unit and fishing in it with one or two other fishermen, so he was not a poor fisher. However, his income has recently become irregular, and his livelihood is in peril, he said:

> We regularly go fishing. But there's no guarantee of enough fish to feed our family. We have to rely on moneylenders/neighbors/relatives if we do not catch enough fish. During potential fish-catch seasons, we are supposed to repay the debt incurred during the rough and fish-ban periods. Our lives have become "cyclic" around debts and moneylenders. During the past five years, we've returned from the sea many times without catching enough fish, and these trends have been rising. I often return from the sea empty-handed. Moreover, we small-scale fishers often return from the sea due to rough weather. Our fishing days have decreased in recent decades. The main victims are fishing laborers and fisherwomen who have no other choice.

On the contrary, a senior fisherman from Chinnamedu described how the remittances sent by his son helped his family to deal with the losses caused by the 2015 heavy rains.

> I own an FRP-OBM unit. My fishing gear and vessel were heavily ruined. Had I visited the coast during heavy rains, I could have saved my gear,

but that was risky. I didn't take that risk and left my gear on the coast. I knew receiving fair compensation from the state would be tough. My son's remittances from abroad helped me make the decision.

A few fishermen respondents also shared this viewpoint and lamented the increasing trends of livelihood crises over the years. Compared to fishermen, who only have catamaran boats and risk their lives in dangerous weather to protect their fishing assets, their financial situation is more stable. Fishermen rely heavily on fishing, especially those using FRP catamarans and FRP-OBM units. To feed their family, they must fish daily. Multiple factors, including erratic and unpredictable weather patterns and climate-related events, reduce fishing days. In rough seasons, fishermen's vessels are increasingly damaged, increasing their vulnerability. Due to inflation, fishermen's vulnerability has risen in the last decade. Many respondents repeatedly reported repairing and renewing fishing assets costs more than their fishing income. Hence, inadequate financial capital has made fishers more vulnerable to climate risks. However, as fishers benefit from state social welfare programs, their food security is not at risk.

Climate vulnerability of surukku valai fishers

Madavamedu and Kottaimedu fishers and a significant proportion of Koozhayar fishers have successfully ventured into the surukku valai fishery. It provides sustainable income to fishing households but is unsustainable for marine resources. Nonetheless, discussions with the respondents of Madavamedu, Koozhayar, and Kottaimedu reveal that they are vulnerable to the effects of climate change and extremes in diverse ways. They believe their fishing days have decreased significantly over the past two decades because of erratic and turbulent weather patterns, resulting in their dependence on moneylenders. A senior fisherman in Koozhayar with over 20 years of experience recently ventured into surukku valai fishing by forming a collaborative partnership with his village's fishers. He stated that it provides a more stable income compared to traditional fishing, though it is not guaranteed. He stated how their ability to make a living from fishing has been hindered by unusual weather patterns, such as the unpredictability of recent large tides, and why most fishermen in his village can only engage in small-scale or near-shore fishing.

> We must enter the interior sea to catch commercial fish, which costs more fuel and fish workers. Deep-sea fishing without safety precautions is risky due to unpredictable weather. As they take up too much space, we don't carry sea-safety precautions when fishing. We received free life jackets as part of tsunami rehabilitation. After that, we heard nothing from the state or external actors. It has been more than ten years, so it is outdated. Most of us do not use modern scientific equipment to identify the sea's direction or fish shoals though we sometimes carry GPS for navigating vessels. For all these reasons, we fishermen do not like taking risks, so we usually fish close to shore.

Figure 5.3 Fishing vessels. Place: Madavamedu.
Source: Picture credit: Devendraraj Madhanagopal.

Madavamedu and Kottaimedu fishers are predominately involved in surukku valai fishing. So, the field observations in the fishing households of these villages showed that they are better off than those in other villages. According to interviews and focus group discussions with Madavamedu and Kottaimedu fishermen, it was found their cohesion and capacity for self-organization are significantly superior to those of other study villages; hence, they collectively invest in surukku valai fishing. Profits and losses are shared without conflict. These findings partially support Kalikoski et al. (2010), who write that fisher self-organization reduces adverse weather vulnerability. Contrary to this, based on the findings, this book highlights that the surukku valai fishermen are likewise vulnerable to climatic and weather stressors though they are financially stable. Overfishing by trawlers and erratic, unpredictable weather patterns undermine their adaptive capacity. Most of them, despite this, are satisfied with their current earnings, even though they are aware that this is not sustainable, and in the long term, they need to face challenges. Furthermore, they bemoaned the fact that while the price of labor and diesel have risen in recent years, the price of fish has not. For example, a surukku valai fisherman in Madavamedu explains why their livelihoods are so vulnerable. One of the youngest uur panchayat leaders, he has worked as a fisherman for 15 years.

> Our fish prices do not cover our expenses. In recent years, all our expenses have risen, including diesel, labor, family costs, children's education, fishing vessels, fishing nets, etc. We frequently take on debts from moneylenders and exchange funds when we catch a sizeable fish. The unpredictability of the weather conditions have decreased our fishing days over the years, and fishing is becoming increasingly risky. Recently,

my boat was capsized by high tides. Somehow, I reached the shore. Relatives and villagers helped me bring the boat to shore. However, I endured a heavy financial loss and an injury. We fish carefully in rough seasons to avoid such accidents. During normal times, we sometimes go fishing based on what we hear on social networks about what fish are available. Such accidents in normal seasons are partly due to fishers' lack of alertness and mostly due to erratic weather. Since the tsunami, such trends have grown.

Surukku valai fishers across the study regions have made a similar point. The analysis shows that all fishers are vulnerable to climate change, regardless of socioeconomic status. Whereas their adaptive capacity varies on their financial, social, and institutional capitals.

Shrinking coastal spaces and raising vulnerability of marine fishers

Fishers showed limited interest in being involved in planned localized adaptation actions to climate change. The only possible way for small fishers to convey their pledges and grievances to the eyes of governmental authorities is through uur panchayats. For example, all the fisher respondents viewed coastal erosion effects as having started to threaten their livelihoods and

Figure 5.4 Surukku valai fishing nets. Place: Kottaimedu.

Source: Picture credit: Devendraraj Madhanagopal.

fishing spaces. However, fishers' capacity is too low to make planned adaptation efforts to respond to erosion. In Koozhayar and Madavamedu, the distance between the coast and the two parts of this village is around 1 km and 750 meters, respectively. Such substantial space between the beach and their house settlements has made them care less about responding to erosion. I repeatedly observed that the respondents, irrespective of their socio-economic category, show limited emphasis on pro-active local climate actions unless the troubles are immediate. In Chinnamedu, Chavadikuppam, and Madavamedu, the fishers perceive that the shore erosion issues are immediate and act in different ways along the lines of their capitals. Houses in Chinnamedu and Chavadikuppam start at around 300 meters from the coast, which makes fishers more prone to climate extremes and shore erosion than their counterparts. Fishers from these villages shared that such a situation has worsened over the past five years (see Figure 5.5). Respondents shared similar thoughts that they are losing their coastal spaces for boarding their boats and doing allied activities like fishing. Nonetheless, discussions with the local leaders of small-fishing villages showed that collective actions to confront coastal erosion have been limited and are highly formal. At most, they conducted a few uur panchayat meetings to fix these issues. Some of their uur panchayat leaders have made efforts to take such matters to the concern of the state administration by giving petitions. Hence, as this book demonstrates, the customary local institutions of the small fishing villages act formally to respond to climate change, which is contradictory. From their accounts, it was clear that they are of perspectives that adaptation actions to confront huge challenges like coastal erosion are beyond their power and capacity.

Figure 5.5 Shrinking coastal spaces. Place: Chinnamedu.
Source: Picture credit: Devendraraj Madhanagopal.

Coping and adaptation strategies of marine fishers

Even though some localized actions of the fishers are not intended directly to respond to climate change and risks, they indirectly serve as coping or adaptive measures. Fishers make many adjustments to weather and climate patterns, including extreme events. Borrowing money from other sources (moneylenders, relatives, etc.) is the primary coping strategy of fishers during rough seasons, cyclone periods, less/small fish catch, and fish ban season. Some coping and adaptation measures/strategies of fishers to climate change and risks are as follows:

- Making more fishing trips in the sea to catch fish than they regularly used to make. Traveling more distance[4] in the sea by using their fishing vessels to catch fish, including commercial fish species like *kola meen*.
- Investing extra money to purchase fishing vessels and equipment.
- Diversifications in fishing operations, such as altering the fishing gears, the simultaneous use of different fishing gears, and the use of varying mesh sizes within the same net, as well as changing the fishing vessels according to the seasons and weather conditions (Salagrama and Koriya, 2008).
- Cut down on costs by not hiring fish workers.
- The deployment of new fishing technologies is financed through moneylender loans.
- During difficult seasons and times of fewer or no fish catches, they borrow money from neighbors, relatives, and friends to support their families. The social networks of fishers are evidenced by their reliance on short-term loans from family members.
- Utilizing the reserve funds (small fishers rarely save money).
- Selling the jewelry[5] of the female members of the households.
- Attempting foreign job opportunities via fisher's social networks (migration).
- Fisherwomen don't have many ways to get money besides borrowing it from family or a moneylender.
- Women's self-help groups (SHGs) are the only viable option to meet their economic needs. Fisherwomen receive loans (microfinance) from SHGs for family support and expenses, but not for their own needs.
- Utilizing the state's social welfare programs, such as the pensions for the elderly, widows, and the disabled.
- Public distribution is Tamil Nadu's most successful government program in recent decades. Throughout the year, fishing populations depend heavily on this program. In the fishing villages, the fishermen cooperative society acts as an effective intermediary in delivering the benefits of this program to the fishermen (*Government of Tamil Nadu*, 2022a).
- The Fishermen cooperative society plays a significant role in coping with disasters, lean times, and rough periods, but not the fisherwomen cooperative society, which is also among the local institutions. Through the

voices of local fishermen and fisherwomen, as this book argues, across the Coromandel coast of Tamil Nadu, fisherwomen cooperative society, exists only in name, as they lack significant roles, resources, and influence among fishermen to accomplish their obligations.
- During the period of the fish ban, the state government provided financial support (formal credit) – The contribution made by the government to meet the economic needs of fishers during lean times.
- Most fishers have "insurance" policies for their fishing vessels. Importantly, the marine fishers who do surkku valai fishing are aware of life insurance[6] options and pay the premium regularly. Life insurance is unknown among Chinnakottaimedu and Chinnanamedu fishers. Their income patterns are always unknown; thus, they can't pay the premium. Few fishers have turned to agriculture or livestock farming in all study villages. Twenty to thirty fishing households in Koozhayar rely on this occupation. Both elected local bodies and uur panchayats implement MGNREGA efficiently. It's a main source of income for 30–50 days a year. DevendraBabu et al. (2011) investigate informal village panchayats in Karnataka's MGNREGA implementation. Like this, recent studies highlight MGNREGA's potential to enhance community climate resilience and call for incorporating social protection programs into the climate resilience framework (Adam, 2015; Godfrey-Wood & Flower, 2018). Nonetheless, currently, no exclusive work exists to discuss how the MGNREGA program has helped Tamil Nadu's marine fishermen adapt to climate change. MGNREGA provides a lifeline for small-scale fishers during rough seasons and fish ban periods, particularly women. So, this plan must be regularly examined and reframed to fulfil small-scale fishers' climate adaptation demands. A few external organizations, notably the M.S. Swaminathan Foundation and the Reliance Foundation, provide weather-related information to the fishermen in a few study communities. However, these organizations do not provide information on potential fish stocks, shoals, etc.

Migration scenario in the Pattinavar fishermen

By 2050, 50 to nearly 200 million people will have to migrate due to global environmental change and climate stressors like floods, droughts, and sea level rise (Warner, 2010). However, climate change is not the only cause of human migration; several social and political factors are the major causes worldwide. Still, empirical evidence confirms that environmental changes, including climate change and environmental hazards, may increase the rate of mass human migration. People move across countries, states, and regions to avoid the worst climatic stressors and live in risky areas (Jäger et al., 2009; Warner et al., 2009; Warner, 2010). As indicated by numerous global studies, including one conducted in India (Salagrama & Koriya, 2008; Vivekanandan, 2010; Islam, 2013; Islam et al., 2014c), migration is a viable and readily accessible option. The migration does not always indicate vulnerability. Warner

(2010) states there are many argumentative views on migration and whether it should be accepted adaptation as a part of it or considered a failure of adaptation. Bogardi and Warner (2009) write migration can also be considered a form of adaptation. According to interviews, the fishermen are more interested in taking jobs abroad by using their social networks. The head of the Koozhayar fishermen cooperative society explained why more fishermen seek employment opportunities abroad. He was one of the key informants. He has been an active fisherman for the past quarter-century and is now engaged in fish trading. He stated,

> This generation's fishers live better than earlier ones. We have TVs and cell phones, and some houses have computers. Nonetheless, as small-scale fishers, we lag behind nearby big fishing villages in many ways. We just began surukku valai fishing, and this income is inadequate to meet our expenses. Since mechanized fishermen have depleted the sea, we do not think investing more in fishing is wise. Small fishers catch what they leave out. Thus, fishing has become a difficult and uncertain, and it is no longer as easy as outsiders believe. Climate change has also increased the risk in our fishing occupation in recent years. No one knows whether we will return to our home safely after fishing, and this fear continues in our lives almost every day. Small-scale fishermen cannot support their families today, and we do not want future generations to suffer. Therefore, we want our children to leave this ancestral profession by getting a good education. We are primarily interested in overseas employment opportunities, which is the best way to make and save money.

Focus group discussions with the fishers and the local key leaders on job migration showed that they are generally not interested in pursuing local-seasonal jobs during the lean period. The government provides certain financial assistance (INR 5000)[7] to fishing families (every ration card) during the lean periods. Fishing households partially fund this amount. From January to August, they pay INR 150 to the fishermen cooperative society. Like this, the women pay a small amount to the fisherwomen cooperative society. In October, the government compensates fishermen during the lean season by distributing INR 5000 to fishing families through the fishermen cooperative society. The state government has started giving fishing households INR 5000 (per ration card) during the fish ban. Interviews and focus group discussions indicate that most fishermen are dissatisfied with this amount since they perceive it as insufficient to support their families in an era of inflation. However, fishers are not interested in seasonal or manual occupations in their villages or adjacent towns. A few fishermen complained that taking up seasonal employment other than fishing is an insult to their livelihood.

According to fishermen, a work overseas is the best way to secure and grow their family's income. Non-participant observations reveal that migration gives income security to fishers, except in a few circumstances. Small fishers of Chinnamedu and Chavadikuppam villages blame overfishing, fewer fish catch, fish species decline, erratic weather patterns, and erosion for their seasonal migration to foreign countries. However, the fact that fishers' seasonal migration and their interest in foreign job opportunities is not a new phenomenon. It has various other cogent reasons, such as less fish catches over the years and family reasons. Interestingly, fishermen do not choose local jobs and livelihood diversification activities in their native villages. They are not willing to take on any local jobs during the rainy seasons, cyclones, or the fish-ban period. It should also be noted that only fishing households with manageable credit and wide social networks can work overseas. This is not limited to the current situation. Asiatic history from the previous century confirms this (Amrith, 2013). Social networks and family credit are inextricably linked to seizing opportunities abroad. Overall, eight months per year are the maximum number of fishing days for fishermen. In addition, the fishing villages that this study investigates lack space and infrastructure to store, dry, and process fish during lean and cyclone months (Kodiyampalayam and Madavamedu are exceptions).

To combat climate change and livelihoods loss, should fishermen migrate?

The household income of fishers is significantly linked to their fishing vessel and gear ownership, indicating fishing assets provide financial security. Fishers who do not own vessels mostly wish to work abroad. Fisherwomen encourage their men to work abroad temporarily, claiming that by working abroad, fishermen can save money without "squandering" it, which improves the well-being of their families. By citing previous research (Keck & Sakdapolrak, 2013), I argue that migration overseas (seasonal and out-migrations) of small-scale fishers is one of the potential coping strategies that can increase the resilience of fishing households to climate change and risks. The results of this book on the vulnerability and adaptation options of marine fishers support earlier discussions and findings on fisheries (Janakarajan, 2007; ICOR, 2011; Black et al., 2011; Islam, 2013; Islam et al., 2014a, 2014b). This book also highlights that migrant fishermen ensured food security and the economic capital of their households. It reflects the increasing adaptive capacity of fishing households to respond to climate stress. State institutions should encourage fishers' livelihood diversification and occupational mobility, says Janakarajan (2007). In the context of the Coromandel Coast of South India, migration is similar to occupational mobility. Fishers hesitate to engage in shrimp farming, poultry and animal husbandry, and other fishing occupations. The state administration can work with local institutions in such an environment to provide better education and alternative employment opportunities for fishing communities. It can help fishers adapt to multiple stressors, including climate change. Discussions with local leaders suggest that many senior and middle-aged fishers have a basic education, and most young and middle-aged fishers have 0–8 years of education. It means the majority of them can read and write Tamil. Fishers generally rely on scientific information to protect themselves from climate extremes. Anecdotal evidence from this study shows that the fishers' educational levels are too low to understand scientific information about climate extremes, which hinders their adaptive capacity.

Further, as discussed in the previous chapter (Chapter 4), younger generations are disconnecting from their traditional indigenous knowledge. Insufficient education levels of fishers to understand climate extremes have reduced their grip on indigenous knowledge systems, making them more vulnerable to climate change and reducing their adaptive capacity (Sharma et al., 2013). Past hazard experience, evacuation experience, and shelter quality influence fishers' response to disaster warnings. Coastal evacuation in India is dominated by community leadership and decisions (Sharma & Patt, 2012; Sharma et al., 2013). According to a recent Central Marine Fisheries Research Institute census, fishing communities in Tamil Nadu are disadvantaged in education (CMFRI, 2012b). The respondents, on many occasions, reported that, in general, fishermen do not encourage their children to pursue this precarious occupation. The fishers want to educate their kids for their well-being. However, children are raised in a social atmosphere. Fishing and its allied activities provide a natural environment for the male children of fishers. So, fishers know it's hard to disconnect their children from "fishing." Moreover, many of these children began fishing with their fathers and relatives very young, particularly during vacations. Due to this, they lose interest in education after a certain period. Chinnamedu, Chavadikuppam, and Chinnakottaimedu have more than 15 college graduates. In Chinnamedu and Chavadikuppam, I saw many young people with diplomas working as co-fishers on their family's fishing boats. They told me they could land jobs because of their education, but the pay was so low that they had to leave and return. Fishers sometimes drop their children from higher education due to low fish catches and erratic weather patterns, which is not uncommon across this coastal stretch.

Disaster risk reduction capacity of marine fishers

For disaster preparedness and post-disaster recovery efforts, it is necessary to identify vulnerable regions and the factors that exacerbate social vulnerability (Cutter & Finch, 2008). Parthasarathy (2015) notes that disasters are about how institutions function, react to risks and have strategies. The study's findings support the two claims mentioned above. The disaster risk reduction capacity of the fishers in the study area is low because their institutions aren't strategizing. Small fishing villages lack the resources/team to respond to environmental disasters, and they have fewer and scattered dialogues with the policymakers and state. This book clearly warns that the fisherwomen and children along the Coromandel coast will be more vulnerable to climate events and disasters. For instance, Chinnamedu, Chavadikuppam, Chinnakottaimedu, and Kottaimedu have no cyclone shelters and lack storerooms for precautionary safety measures and emergency equipment to face climate events and natural disasters, making them more vulnerable to disasters. Interviews and discussions with local leaders and senior fishermen of these villages suggest that fishers show the least interest in evacuating cyclone shelters in response to state cyclone warnings. This is due to a few factors. First, fishers are reluctant to evacuate to protect their physical and fishing assets. Second, as previously mentioned, the fishers know that cyclone shelters lack adequate space to accommodate the entire village.

Barriers and limitations to coping and adaptation efforts of marine fishers in responding to climate change

Moser and Ekstrom (2010) define climate change adaptation barriers as "impediments, that can stop, delay or divert the adaptation process, or that might prevent the community from using its resources most advantageously to respond to climate change impacts." Several authors (Adger et al., 2009; Nielsen & Reenberg, 2010; Moser & Ekstrom, 2010; Jones, 2010; Islam et al., 2014b; Kuruppu & Willie, 2015) discuss the limits and barriers to climate adaptation in different settings. This study largely confirms earlier findings. In addition, based on field findings, a few new limitations have been identified, such as the inability of the Pattinavar fishers to plan and execute adaptation actions to climate change. These insights provide certain climate change lessons that can be matched with the other coasts around the world.

Cultural and social barriers[8]

Male fishers dominate the fishing and uur panchayat, restricting women's participation in institutional activities in fishing villages. Uur panchayats are all-male panchayats. Women do not even attend ur panchayat meetings. Fishing has been a "men's occupation" for generations. Fishermen's dominance in the fishery has historical and cultural roots. Because of this, they have not been able to think about collaborating with the fishermen to create collective responses to the multiple stressors, including climate change, that they face. Hence, this research identifies those fisherwomen who have a limited understanding of climate change effects on fisheries. In coastal Tamil Nadu, fisherwomen have a separate cooperative society called "*Meenava Pengal Kootturavu Amiappul Sangam.*" Nonetheless, its roles and activities have not been well-structured throughout the study area. Hence, cultural barriers of the fishing villages evidently limit the adaptive capacity of Pattinavars to climate change.

Communication and technological Barriers

Jones (2010) discusses how technological, and resource (including financial) limitations can hinder climate change adaptation. Fishers are dependent on the district fishermen cooperative society for early warning alerts. In the event that fishermen require urgent weather alerts, the employees of the district administration directly visit those vulnerable fishing villages and alert fishers. Nevertheless, as senior fishers noted, such alerts are only meant for emergencies. Small fishermen in the research area have trouble identifying fishing zones. Their fishing assets are often devastated by severe winds and high seas. Due to the unpredictability and unpredictability of the sea's weather, they frequently confront life-threatening dangers. As a result, they must adjust their fishing schedule to deal with unpredictable weather. Due to a lack of accessibility to technology, fishermen face communication constraints that limit their climate change adaptation strategies. Since the tsunami, as I highlighted in the last chapter (Chapter 4), fishermen have begun to realize that their local knowledge systems are insufficient and often inaccurate. In this situation, fishermen need accurate weather reports as soon as possible, which most do not have. In recent years, due to the development of information technology, fishermen have been regularly updated and warned about cyclones and storms, despite their belief that only apparent warnings are issued at evident times (Janakarajan, 2007). Such complications arose during the massive downpour that swept these coastal regions in 2015. The refusal to seek shelter in cyclone shelters increases their risks, although they have not suffered significant damage in the last decade. Fishermen in the study area generally don't carry safety equipment while fishing. This lack of safety knowledge also increases their vulnerability.

The state's apathy

Pattinavars are less likely to be socially excluded in their fishing communities since their social networks and self-organization mechanisms supply them with a significant amount of social capital. Therefore, they never received any chauvinist threats from external forces. But fishers, especially small fishing villages, felt socially left out of state institutions and bureaucratic machinery. Fishing villages are geographically remote from towns. For example, Chinnamedu, Chavadikuppam, and Kodiyampalayam fishers live near the sea, making them more vulnerable to shifting weather patterns and climate risks. They believe that the town and district governments do not care much about their lives and livelihoods and feel "alienated." Every rough season, rainwater remains stagnant for a week after the rains recede, causing minor or severe health issues. However, as they pointed out, local officials and elected politicians rarely visit their villages to learn about their health. The only time elected politicians go to their fishing villages is when there is a weather event. They noted that apathy toward small-scale fisheries is common in the governmental apparatus, regardless of political orientation. In focus groups, local leaders often raised the issue that small-scale fishers were not compensated when their fishing assets were damaged during rough seasons. They mentioned the poor relief measures they received after the 2011 Thane cyclone. One of Chavadikuppam's key informants said:

> After climate extremes, government officials undertake surveys and offer various guarantees. However, only a few of their promises have been kept.

In all the fishing villages, particularly in Chavadikuppam, Chinnamedu, and Chinnakottaimedu, the same sentiments were expressed. Marginalization heightens the climatic vulnerability of impoverished fishermen. Also, senior fishermen and local leaders of small fishing villages, such as Chinnamedu and Chavadikuppam, believe that local NGOs primarily support large fishing villages involved in capital-intensive fishing. A lack of revenue and individual and collective economic capital meant they could not negotiate effectively with the state administration. As a result, their concerns go unheard and unaddressed. As stated in prior chapters, the fishing villages along the coast of Tamil Nadu are predominantly homogeneous. As a result, local and regional social institutions have historically wielded considerable power in influencing the climate change vulnerability of Tamil maritime fishing communities. Consequently, their fragile economic conditions and the indifference of the state administration are among the few elements that exacerbate their vulnerability to climate change and risks. A few additional factors, such as the economic capital and political networks of fishing communities, also play a key role in shaping and generating the vulnerability of coastal fishers.

What enables marine fishers to adapt effectively to climate change?

While highlighting the factors that increase Pattinavars' vulnerability, this chapter explores the coping and adaptation mechanisms of fishermen to climate change by highlighting a few typical narratives from the field. Many conclusions of this chapter mirror ICOR's (2011) conclusions on coastal Mumbai. Like this, the localized coping and adaptation actions of Pattinavars are widely echoed by the findings of previous studies conducted in many contexts throughout the world (Janakarajan, 2007; Badjeck, 2008; Salagrama & Koriya, 2008; ICOR, 2011; Islam, 2013; Marschke et al., 2014). This book also has certain unique findings, focusing on the Coromandel coast of Tamil Nadu, which is relevant to other coasts of India and around the world. This study demonstrates that ownership of fishing assets, economic capital, and political capital play determining roles in deciding the coping and adaptive capacity of fishers in responding to climate change and risks. This is match those of a recent study (Rotberg, 2012) that identifies that poverty exacerbates the financial insecurity of fishermen and, as a result, impedes their adaptation actions to climate change. Social networks and social capital can help fishers cope with climate impacts and lower their vulnerabilities without external support. For instance, in Chavadikuppam and Madavamedu, such mutual aid and social networks of fishermen may have been useful in 2015 in coping with high rains. Formal and informal networks of trust and social support can help fishermen deal with climate change in the near term, but they cannot do much to strengthen their long-term adaptive ability and community resilience.

Compared to other study villages, the fishing households in Madavamedu and Kottaimedu are more prosperous. Here, cultural standards were rigidly against women. Discussions with the fishermen of these villages demonstrate they widely held the belief that women's economic engagement is unnecessary. Likewise, fisherwomen in villages are also uninterested in livelihood activities or village affairs. They are oblivious to the existence of the problems mentioned earlier. Even though they have a higher level of education than other small fishing villages, my conversations with them indicate that they are unaware of climate change in their livelihoods and coastal regions. In contrast, the images of little fishing villages showed the exact opposite. For example, Chinnamedu fisherwomen are active members of women's SGHs. Discussions with them indicated that they are largely aware of climate change implications on their livelihoods, including the erosion of their coastlines. From this, we can conclude that poor fisherwomen are ready to challenge their cultural norms and actively engage in income-generating activities. Poverty gave the fishing families a chance to defy the stringent cultural standards of the coastal villages they lived in. For example, the fisherwomen in Koozhayar and Chinnamedu are actively involved in SHG operations, and they possess more understanding of the growing threats of climate change and risks. After the 2004 tsunami tragedy, the state government realized the importance of awareness of disaster risk reduction. It launched efforts to develop a disaster management committee (by local fishermen, especially young fishermen) in all coastal districts of Tamil Nadu. At the district collectorate, the Nagapattinam district has its own disaster management authority. Pre-disaster management planning and climate adaptation initiatives in fishing villages, however, have been restricted by a lack of interconnection across institutions. Continuous and ground-level disaster management plans, grassroots awareness campaigns on "dos and don'ts," training on evacuation and rescue measures during disasters and climatic extremes such as floods, provision of better infrastructure, and identifying and recognizing the social institutions that represent diverse groups in the community are all required to initiate and build adaptation at the local levels. We have sufficient evidence from the Indian state of Odisha to demonstrate that the state's awareness campaigns and effective grassroots awareness initiatives have significantly reduced regional vulnerability and created a disaster- and climate-resilient community. In addition, these programs could positively impact the community's behavior and aid in adapting households to climate change. There is still a need for greater research into awareness efforts, planned disaster education programs, and the media to reduce people's vulnerability to climate change threats (Das & Smith, 2012; Das, 2016).

Notes

1 It is an approximate figure. Throughout this chapter, I point out the socio-economic details of the marine fishers from the data collected from the fieldwork.
2 It refers to the types (local names: In Tamil) of fishing gears in mechanized fishing.
3 To know the latest details, we can refer to the official website of the Tamil Nadu Fishermen Welfare Board. https://tnfwb.tn.gov.in/index.php/User/login (Tamil Nadu Fishermen Welfare Board, n.d.)

4 Season, weather, age groups, and personal needs determine how far fishermen travel to fish. Fishermen informed that they could get enough fish catch and commercial species only if they travel more than 15 m in the sea. Sometimes, fishermen go out in bad weather because they must fish daily to support their families.
5 Thomas et al. (2010, p. 41) discuss the coping strategies of the poor and list the various coping strategies during shocks and stresses. To understand how fishers cope, it's important to look at this study's findings, even though it was conducted in an entirely different context.
6 Bhattamishra and Barrett (2010) present a comprehensive analysis of community-based risk management arrangements, focusing on poor countries. In their research, they distinguish "insurance" – primarily informal mutual insurance and insurance for major life events – from savings and credit arrangements to ensure against income risk (it includes microfinance, cereal banks, grain banks, and other social assistance facilities), which are more institutionalized and often confined by rules and regulations (Bhattamishra & Barrett, 2008; Bhattamishra & Barrett, 2010).
7 Fishers' oral statements about the lean and ban period assistance and the amount they regularly pay to the fisheries cooperative society were unclear. Most women I interviewed were unaware of such details, which showed fisherwomen were excluded from public assets and finance. The numbers I provide in this chapter come from the respondents' statements. For official numbers, we can refer to the latest policy note (*Government of Tamil Nadu*, 2022a) and the official website of the state government of Tamil Nadu. It provides a comprehensive account of State and Central Government welfare schemes for the fishers that the Department of Fisheries is implementing https://www.fisheries.tn.gov.in/WelfareSchemes (Department of Fisheries and Fishermen Welfare, n.d.).
8 Chapter 7 provides an extensive discussion on cultural barriers, gender bias, and women's agency in relation to climate change and risks.

6 Local institutions

Boon or bane in local adaptation to climate change?

Local institutions along Tamil Nadu's Coromandel Coast: underexamined, undervalued, or overvalued?

> Pattinavar fishermen have strong social networks, which we lack. Their lives are connected with the water and coast; thus, they have intrinsic fishing abilities and knowledge. As you can see, they have an emotional tie to fishing while we do not. We see fishing as a means of meeting our financial demands. We are happy to try any other job that pays well. In this case, what is the point of participating in "Uur panchayat" activities? It takes a lot of time.

This is what a fish worker from the village of Koozhayar said when I questioned his engagement in community activities through local institutions. He stated with a grin that he was not from a fishing caste. However, for the past two decades, he and his family have subsisted solely on marine fishing. Despite this, his participation in uur panchayat activities has been minimal. Such stories are not uncommon in diverse (or non-homogenous) fishing villages. All respondents of this study, regardless of socioeconomic status, agreed that "climate change" is real and brunt; however, their interpretation is based on experiential and local knowledge, not science. Local institutions in coastal India have received limited attention in social science academics and climate change literature, and they are mostly underexamined, resulting in either oversimplification and underestimating or overestimation of their capabilities and influence on fishers and coastal regions. This insufficient and erroneous understanding of local institutions places already vulnerable communities in more precarious positions in combating climate change. To explain this more, (i) given their limited understanding of the local institutions, the state and external actors of coastal India overvalue the strength of the local institutions (can say, uur panchayats) and offer greater assistance to them in reacting to climate change without considering the disparities among the local institutions and the character of the fishing villages to which they belong. (ii) An overemphasis on uur panchayats may also cause the state and other external players to overlook the potential of other institutions, such as

DOI: 10.4324/9781003187691-6

newly growing local institutions like women's self-help groups (SHGs). It also reduces the resilience and adaptive capacity of fishers to climate change. This chapter urges readers to acknowledge that climate change is a global issue with numerous local ramifications. The actions and capacity of local communities to adapt to climate change depend on the complex interaction of their "capitals." Investigating the roles, functions, potentials, and constraints of local institutions in coastal areas can reveal fresh and novel answers as well as solutions to climate change. In order to effectively respond to unforeseen concerns such as climate change and risk, it is necessary to examine not just the activities of powerful local institutions but also those of less influential local institutions.

Recalling the conceptual talks on local institutions in this book's Chapter 2, I bring Gupta et al. (2010)'s definition of institutions into this context: "Institutions reflect formal governmental processes as well as formal and informal social patterns of engagement." This definition encompasses both formal and informal institutions. Again, recalling the academic discussions on local institutions presented in the Chapter 2 of this book (for example, North, 1991; Helmke & Levitsky, 2003; Agrawal, 2008), this chapter provides an overview of the primary actors, roles, responsibilities, and social networking capacities of the local institutions across the research region. In doing so, it emphasizes uur panchayats, the prominent local institutions along this coastline region. Following that, it investigates the scope of local institutions in responding to climate change using empirical insights. It illustrates how responses of local institutions in fishing villages differed dependent on their political and financial capital. It then discusses the limitations of local institutions, highlighting the antiquated characteristics of local institutions that hinder their ability to adjust locally to climate change. This chapter includes climate change works from many regions of the world throughout these discussions, supplementing and balancing the scholarly discourse.

Institutions for marine fisheries in India

Small-scale fishing communities have historically managed their fishing resources in many regions of Asia, resulting in the development of several local associations, family and clan groups, traditional institutions, traditional leadership, and individual groups over time. Social science scholarships in recent years have explored and investigated several facets of fishing communities and coastal systems. Some select works/areas include Natural Resource Protection, Long-Term Sustainability, Social Capital (Pretty & Ward, 2001; Andersson & Agrawal, 2011), Strengthening Fisheries Governance (Bavinck, 2001a, b; Jentoft, 2004; Pascual-Fernández et al., 2005; Chuenpagdee & Song, 2012; Novak & Axelrod, 2016), Capitalism and Degradation of Marine Ecosystems (Clark & Clausen, 2008), *The Tragedy of the Commodity* (Longo et al., 2015), Marine Sociology (Longo & Clark, 2016), *Too Big To Ignore* (Chuenpagdee, 2019), Small-Scale Fisheries Governance (Chuenpagdee & Jentoft, 2015 Chuenpagdee & Jentoft, 2018). In the fisheries sector in India,

112 *Local institutions*

institutions and agencies include a wide range of diverse stakeholders, which is growing due to new players and processes (FIMSUL, 2011). The recent FIMSUL (2011) report categorizes the formal and informal institutions essential to fisheries management and governance. They are as follows: (i) State fisheries department, (ii) Union Ministry of Agriculture (MoA), (iii) Union Ministry of Environment, Forest and Climate Change (MoEF), (iv) State forest department, (v) Central Marine Fisheries Research Institute (CMFRI), (vi) fishermen traditional institutions (customary local institutions), and (vii) mechanized boat institutions. These institutions organize roles and powers to carry out the laws (including community laws) governing fisheries systems. The social and cultural characteristics of Indian marine fishers have been tied to their coasts for generations. On the Coromandel coasts of Tamil Nadu and Puducherry, local institutions, particularly uur panchayats (fishermen councils), play a vital role in fisheries management, legal pluralism, and the sharing of sea territories. Marine fisheries systems of coastal India are complex and are composed of various institutions, actors, and interplays among the institutions (Kurien, 2020; Allison et al., 2020; Bavinck, 2020; Baiju, 2022). As the effects of climate change continue to worsen, the multiple ranges of institutions of marine fishers across coastal India must implement both short-term and long-term adaptation actions.

Exploring local institutions: types, functions, and critical perspectives

Patttinavar' uur panchayats[1] – the most influential local institution

Caste and religious committees have active control over the fishing communities of Tamil Nadu's coast. Fishermen panchayats exclude women and other fishing castes from managing the village's resources (Aldrich, 2011), as detailed in the next chapter (Chapter 7). As delineated by Bavinck (1998), the primary responsibility of uur panchayats is to regulate the fishing rights of fishermen in fishing villages and to prohibit the use of destructive fishing nets. He notes – On the coast, two legal systems coexist. One is a formal legal system emanating from the state, and the other is an informal (non-state) fishermen council (uur panchayats). As he observes:

> State law ends where fishermen law begins and where fishermen law is effective, the state sees little reason to become involved.
> (Bavinck, 1998, p. 165)

Bavinck & Salagrama (2008) examine the fisheries of Tamil Nadu through the lens of interactive governance theory, noting that the mutual understandings among the uur panchayats grant them jurisdiction over a land area in both the inner land and the sea. According to Bavinck (2011) in another work, uur panchayats exert influence over coastal people and resources along the entire coasts, culminating in the sharing of their shared territorial rights. The jurisdiction of the uur panchayat of one coastal village terminates at the boundary of another village, after which the administration of the uur panchayat of the neighboring village begins. Adopting this strategy puts the entire coastline under "uur panchayats."

Functions and governance systems

A typical uur panchayat/naattaar panchayat (in Tamil) of a fishing village in Tamil Nadu and Puducherry's Coromandel Coast consists of around five male members/leaders. This number may vary depending on the geographical size, population, and traditional practices of the village. In practice, as noted earlier, the fishers do not follow any delineated criterion or tactics to select the panchayat leaders. The uur panchayat contains a group of leaders in a fishing hamlet, including the *panchayattar* (a rank lower than the uur panchayat leaders) and *kanakkaalar* (In Tamil: Treasurer). uur panchayat leaders are not chosen through democratic means (universal voting). They are, however, egalitarian, transparent, and accountable to the Pattinavars (Bavinck, 2001a, b). Before a few decades, the fishing villages that dot the Coromandel coast elected the heads of their fishermen councils using lineage systems. However, the system of selection based on ancestry has become outmoded in recent decades. Villages, on the other hand, have not embraced open democratic election systems in place of lineage systems. They switched to the nomination system to choose representatives (uur panchayat heads) through collective deliberation and collective consent among local fishermen (Bavinck, 2016). In practice, former members and senior fishermen from the village have sway over the selection of panchayat heads. Women are universally excluded from all uur panchayat participation and decision-making processes. There are no specified prerequisites for becoming an uur panchayat leader. Any adult fisherman in the Pattinavar community is eligible to serve for uur panchayat. However, the Pattinavar fishermen's community requires a leader to possess certain qualities. They are as follows: (i) Extensive fishing experience, (ii) effective communication skills, (iii) the ability to interact and negotiate with fishermen and "external" institutions, particularly bureaucrats, and (iv) an understanding of the world outside of fishing. Discussions with Pattinavar fishermen reveal that educational qualifications for uur panchayat leaders are not rigorously considered. Senior fishermen urge junior fishermen to take on the role of uur panchayat head, provided they are educated, engaging, and savvy in dealing with external actors. Though there are some discrepancies in how fishing villages elect their leaders/members of uur panchayats, organizational structures, roles, activities, and governance systems are essentially the same across all fishing villages in this region. As a result, the core governing systems of uur panchayats do not conflict with one another (Bharathi, 1999; Bavinck, 2001a, 2001b; Gomathy, 2006; Swamy, 2011; Swamy, 2021). Scholars who researched the Coromandel Coast note that uur panchayats play crucial and interrelated roles in nearly every part of socioeconomics, culture, and politics in the life of Pattinavar fishers along Nagai – Karaikal coasts and wield exclusive local power over practically all aspects of their existence (Bavinck, 2001a, b; Chandrasekhar, 2010; Novak & Axelrod, 2016). Their primary roles lie in the management of local affairs and resolving their local issues without the interference of outsiders. They play critical roles in organizing local temple festivals, cultural activities, and even the civic affairs of fishermen. Along with that, they also collectively address the fishing rights of small-scale fishers and over-exploitation, and unsustainable fishing issues of trawlers (Gomathy, 2006; Thamizoli & Prabhakar, 2009; FIMSUL, 2011; Bavinck, 2016). They keep an eye on village order and levy a "fine" on those who violate the uur panchayat's set of norms and values. Thus, in fact, uur panchayat leaders at the village level are primarily responsible for resolving local issues and community management but not for directly supporting the marginal fishers for their livelihoods during climate "shocks" and crises. These customary local institutions deal with local issues among the fishers within and across the fishing villages and maintain peace by resolving their disputes. It combats overfishing and ensures fisheries and coastal ecosystem sustainability (Baavinck, 2016). As Esther Fihl writes on the crucial roles of the uur panchayats in the face of resource conflicts among Pattinvar fishers within and among the fishing villages of the Coromandel coast, "…if the caste councils were not there to restrict lives, then disorder, violence and assault would be part and parcel of their life-worlds."

Overall, the customary institutional arrangements of uur panchayats are tightly intertwined with other institutions. Significant past research on the local institutions of fishers in this range is mostly on geography and caste. In addition, uur panchayats have influenced panchayat raj institutions in recent decades (Gomathy, 2006; Bavinck, 2016). They represent the interests of fishermen to the outside world and act as a powerful mediator/negotiator between the state and the fishers, including non-fishing communities in the village. The vital role of uur panchayats in recovery and rehabilitation, linking humanitarian agencies with fishers, was lauded. In addition, they negotiated with the local authority to receive compensation for fishing assets lost by the 2011 Thane cyclone and the 2015 heavy rains. Collaboration between uur panchayats and other local institutions benefits adaptive co-management systems, which help fishers adjust to climate change and extreme events.

Extensive field research demonstrates that uur panchayats lack their sources of income. Therefore, they primarily rely on taxes and fines levied on fishermen and fishing households to fund local temple celebrations. As observed by Bavinck (2016), uur panchayats do not follow specific protocols to collect tax from fishermen; instead, they rely on local circumstances. Uur panchayats collect taxes from adult fishers using state-like procedures (for example, using ration cards to collect taxes). This strategy has been extended to "aal vari" (In Tamil: Person-tax) in recent years – It helps uur panchayats collect taxes from male wage earners and "veettu vari" (house tax). By adopting this method, uur panchayats collect taxes from fishing households based on their ownership of fishing assets. Respondents reported that the source of these membership lists is ration cards of the state. Over the past few decades

Figure 6.1 The typical uur panchayat meeting (Fishers discussed the upcoming temple festival. Place: Kodiyampalayam).

Picture credit: Devendraraj Madhanagopal.

and years, the governance systems (including the methods of collecting taxes) of the uur panchayats have consistently evolved to meet the changing socio-economic transformations of the fishers and the entire state. In this context, I also highlight that the uur panchayat ensures that the welfare schemes of the state reach every fishing household without exemption. This is consistent with Bavinck (2016), who observed that the uur panchayats did not adhere to strict or comparable procedures when collecting from the fishers. In collecting taxes from the fishermen, they are lenient and accommodating. In some early works, Bavinck (2001) highlights how village-level social networks are often robust in uur panchayats and well-established social structures. Regardless of socioeconomic or political differences, the village fishermen unanimously adhere to the uur panchayat's decisions. The same holds for this study. Nonetheless, during the fieldwork in Kodiyampalayam in 2015, I noticed a few verbal confrontations between fishermen and uur panchayat leaders. Local fishermen casually remarked that similar incidents are also uncommon in these fishing villages; however, such disagreements are typically limited only to a certain extent; the senior fishermen and other local heads intervene and ensure that it should not burst into a "big" issue as they always aim for integration of the fishing community.

The transparent and democratic nature of uur panchayats includes this as well. These conflicts are typically resolved peacefully by the village's senior fisherman. Hence, this study reveals that no such significant events in the recent past kindled tensions between villagers and uur panchayats. As a result, the fishermen are unlikely to oppose their panchayats' decisions. Besides this, there is also a practice of inviting respected village elders from surrounding fishing villages to mediate and resolve knotty disagreements among the fishers, ranging from local conflicts to fishing rights. In principle, the social networks and the authority that the uur panchayats wield over the locals are quite strong within the fishing villages. However, such strong social networks and uur panchayat authority at the village level become weak over the shore. It is consistent with Bavinck's (2016) views, in which he stated that the structure of uur panchayats is strong at the basement level (uur, in Tamil). However, the Taluk (Sub-district) lacks a solid institutional structure. On average, there are 11 fishing villages. Bavinck included the entire Nagai-Karaikal coastal region in this category. Basement levels are not as strong as the above ones. He recounts how Akkaraipettai village usurped the power to operate as "thalai gramam" from Nambiar Nagar, which had traditionally been the head of the entire coastal region. This book highlights the three most important barriers to the function of social networks of the uur panchayats: (i) Limited financial capital, (ii) lack of/weak political will, and (iii) failure to create long-term partnerships with the state (Bavinck 2016). Though this study agrees in part with Bavinck's findings, it emphasizes that small-fishing villages continue to struggle to form long-term partnerships with the state, not only due to a limited political will but also due to limited human capital. Besides this, predominant engagement in small-scale fishing practices also results in a local barrier to the fishing villages in responding to climate change (but not surukku valai). It causes most fishers in small fishing villages to live in precarious socioeconomic and political conditions, which subjects them to the triple burden of overfishing (by industrial fishermen), pollution, and climate change effects (including the long-term effects of the 2004 Indian Tsunami).

This book's findings match Gomathy's (2011) studies – she notices the following common qualities of Pattinavar local institutions: shared responsibilities, distributed rights, collaborative governance of the commons, democratic decision-making, and egalitarian nature. Beyond that, the expansion of the power of the uur panchayats is also addressed in this chapter. The influence of the uur panchayats extends to other local institutions (formal, informal, and semi-formal) at the village level. This entails lobbying and negotiations with the state administration, as well as with bureaucrats. The key heads of the uur panchayats implicitly decide who the key actors in other local institutions are. This does not imply, however, that the influence of uur panchayats undermines the electoral processes of local statutory entities. Representatives and leaders in Pattinavar villages are chosen by consensus and discussion because of the strong social and kinship links that exist among them. Senior fishermen and heads of the uur panchayats play vital roles in selecting key actors. These findings are consistent with previous research on community panchayats in rural India (Ananth Pur, 2007a; Ananth Pur, 2007b; Palanithurai & Ragupathy, 2008). In some ways, the conclusions of this study echo those of Bavinck and Vivekanandan (2017), in that uur panchayats ensure the well-being of fishing populations on a variety of levels. In addition, backed with field insights, in the following portions of the book. This chapter also sheds light on the ways in which the uur panchayats lack long-term visionary plans/strategies to sustain fisheries' livelihoods and to confront the challenges of climate change.

Local government (statutory gram panchayats)

Fundamentally opposing ideologies and schools of thought converge on a common ground and value decentralization, and it is often regarded as the key to successful social development (Bardhan, 2002; Mohmand, 2005). There are subtle distinctions between the terms: local democracy and decentralization – even though they are frequently used interchangeably in Indian political discourse. There is also the possibility that decentralization could deepen the local caste- and class-based hierarchical power systems, and it may exacerbate the vulnerability of the marginalized (Government of India, 2007). Critiques against decentralization are also not uncommon (Manor, 2003). The Indian government introduced the landmark 73rd and 74th constitutional amendments in 1992 after going through extensive deliberations and several phases of development. It elevated village panchayats to the third tier of India's political structure and endowed them with several specific provisions for the disadvantaged segments of the population. One of the most frequently praised aspects of this amendment was the reservation of seats for women, particularly for Scheduled Caste and Scheduled Tribal women, and it mandates the state to reserve one-third of the seats for women. It has paved the way for the participation of all marginalized groups in the decision-making processes of local administration, but in many instances, the old power structures of caste, class, and patriarchy have remained unquestioned and powerful (Lele, 2001; Rajasekhar, 2004; Government of India, 2007; Palanithurai & Ragupathy, 2008). During the initial phase of implementation of panchayat raj systems in rural regions, there were several instances in which the new formal local institutions, with their legal standing and categorized responsibilities, resulted in the frequent clash with the already existent informal local institutions (Palanithurai, 2001). The formal local institutions (gram panchayats) exerted a strong impact in areas where traditional panchayats were weak, but disputes and conflict occurred in areas where the roots of traditional panchayats were strong (Palanithurai, 2001). Palanithurai (2008) asserts that the influence of caste panchayats/customary village institutions is greater in villages where caste homogeneity is high. The authority of caste panchayats/customary panchayats varies according to caste, historical context, geography, and economic standing.

118 *Local institutions*

Figure 6.2 Panchayat office. Place: Kodiyampalayam.
Picture credit: Devendraraj Madhanagopal.

Like the rest of the state, the tsunami-hit study villages of the Coromandel Coast elect local representatives through local body elections. Elections are often handled smoothly because of mutual respect and cooperation between local governments and the state. Uur panchayats, in many cases, lend their full support and do not meddle with the local government election process of the state. Consequently, parallel governance systems (formal and customary) coexist to manage the maritime fishing villages along this stretch of coastline. The elected members of the local bodies are mainly responsible for bringing financial resources from the state to the fishing villages. By implementing social welfare schemes of the state for the fishers, they support fishing populations. Like this, by collaborating with influential uur panchayats, they play key roles in numerous during and after extreme climatic conditions and disasters. Thus, parallel (formal and customary) governing systems coexist in the Coromandel Coast of Tamil Nadu.

This study confirms that, in the Coromandel coast of South India, elected statutory bodies at the village levels possess fewer functions, activities, and powers than uur panchayat leaders. As the fieldwork shows, this is particularly high in the case of small fishing villages with low populations. The elected representatives of local bodies, in conjunction with the uur panchayat leaders, serve as a link between the fishing village and the state. During weather and climatic threats, their functions are visible. Across the fishing villages of this coast, there is a lack of sense of disaster risk education and readiness among the elected leaders, making the fishers more vulnerable to climate change and risks. In the focus group discussions, the respondents uniformly stated that any adult fisher could run for and win elections, yet many respondents stated that politics and political association are not in their best interests. They went on to say that politically, and financially powerful fishermen are more likely to run for local office. This is because poor fishermen have fewer prospects of winning elections, although kinship and marriage unite them. Discussions with the local heads of small fishing villages like Chinnamedu and Chavadikuppam revealed a few exceptions, indicating their unanimity toward Pattinavar uur panchayats. In many cases, such as in Kodiyampalayam, the fishermen choose their local-body representatives (ward members) collectively in order to minimize election-related disputes among the populace. Political parties influence and intervene in elections in fishing villages with a substantial public revenue base. The best example is Madavamedu. In comparison to their peers, the leaders of fishing villages engaged primarily in surukku valai fishing have gathered greater political and economic capital. In large fishing villages, the elected members of the local bodies and the heads of the uur panchayats share major local authority. In addition, they are networked with the leaders of the neighboring fishing villages and political leaders of the district, and these sociopolitical networks aid the fishers and their households during climate extremes and hazards. In Madavamedu, for instance, elected members of the local statutory bodies made significant efforts to negotiate with the district administration to bring aid to the victims of the 2011 Thane Cyclone, as compared to their neighbors, who are small fishing villages. However, discussions with local leaders revealed their influence is nowhere near that of trawler fishermen when negotiating with the district administration.

Fishermen cooperative societies

The fisheries cooperative society[2] in Tamil Nadu is a state-owned undertaking that works at the state, district, and village levels. It plays a crucial role in executing the state-sponsored social welfare schemes in the fishing villages by working with uur panchayat leaders. Furthermore, it governs maritime fisheries by granting fishers and their assets membership and identification. It takes the lead in documenting and updating information about fishing villages, fishers, and their fishing assets. During natural disasters in the past, fishermen's cooperative groups played various roles in flood rescue, relief, and recovery activities. Unlike traditional uur panchayats, fishermen cooperative societies are governed by the commissioner of fisheries and the state government. A fishing village typically has one secretary for the fishermen's cooperative association. In principle, the uur panchayat of the village chooses the secretary of this association (a local institution) through mutual consultation with the villagers. In extreme weather conditions, the district fishermen cooperative society gives prior warning information to the village fisherman cooperative societies. The secretary of the fishermen cooperative society must warn local fishermen about upcoming threats and alert them. Therefore, these societies serve critical roles in enforcing the fish ban during summer seasons. Therefore, the fishermen cooperative society serves primarily as a formal liaison between the government and the fishing community. The secretaries of the cooperative society of fishermen at the village level are networked with the secretaries of the surrounding villages. This integrated social network among fishermen's cooperatives spans the entire coast and provides the potential stock of a "social capital" pool to the Pattinavar fishing community. It should also be noted that the fishers, in most climate extreme events, anticipate the fishermen cooperative society to urge the state to effectively administer welfare, relief, and savings programs (fishers' relief fund during lean times). The fishers frequently ask the local fishermen's co-op secretary to voice their worries about the relief fund and "improve" their fishing villages.

Fisherwomen cooperative societies

In 1981, the Tamil Nadu state fisheries department formed the Fisherwomen Extension Service to uplift the socioeconomic status of fisherwomen. It aims to integrate fisherwomen with the ongoing development and welfare schemes of the state (Drewes, 1986). Like fishermen cooperative societies, the fisherwomen cooperative society operates at three levels – state, district, and village. This study finds that the roles, functions, and activities of fisherwomen cooperative societies at the village level are visibly weak across the village levels, and their presence is a namesake. The social networks of village fisherwomen's cooperatives are poor. Furthermore, the fisherwomen cooperative societies lack the necessary data to identify fisherwomen vendors and women involved in allied occupations. Typically, the fisherwomen choose their secretary through consensus. In the interviews, the elected heads of these cooperatives often lamented that the local fishermen heads interfere or dominate the selection process for the leaders of their local institutions, stating that the local fishermen leaders generally do not favor holding elections to determine the representatives of the fisherwomen secretary society, thus rendering this institution ineffective and extinct. At times, the fisherwomen's secretaries desire to be involved in domestic matters, including disaster risk reduction. Nevertheless, it is also equally valid that the female secretaries lack adequate understanding of the requirements and functions of the fisheries systems – this is mainly due to their restricted and confined roles that emanate from the sociocultural norms of the fishing villages.

Political parties

Three major political parties[3] are active along the coast, and they continue to have a stronghold among Tamil Nadu's marine fishers over the past four decades. The political parties are visibly active, mostly during elections. In planning disaster risk reduction initiatives and climate change adaptation, neither the political parties nor their main actors have distinct responsibilities, functions, or resources. They are, nonetheless, well-connected among fisherfolk and political party heads at the district level. As pointed out earlier, the study villages have 2000 people or less; therefore, a few do not have influential political leaders. Besides this, due to the higher population density of the surrounding villages (mostly the villages that are predominately engaged in mechanized fishing) than the nearby small fishing villages (the study villages), the local-body leader often hails from one of those villages. Elections for local bodies, state assemblies, and parliament are held regularly, and fishing populations participate (i.e., vote). Other local institutions, such as uur panchayats, have minimal influence over whether or not fishers vote for a certain political party, and the decision is entirely up to the fishermen – they elect their representatives without the intervention of other local institutions. Influential fishermen from fishing villages are important to all parties. Influential local fishermen are highly important to all the political parties as they are well connected regardless of their political leanings and differences, and such networks are advantageous to the fishing villages in securing funds from external actors (including the state) for the prosperity and "development" of the fishing villages. Since political parties are primarily concerned with "votes," small fishing villages with a limited population receive less attention from political parties than fishing villages with substantial populations. This impacts fishing communities' ability to adapt to changing climatic conditions and threats. For these reasons, this book also contends that the human capital of the fishing villages makes a significant difference in negotiations with political parties and district authorities. In many cases, as fishers shared, the human capital of the fishing villages is equally important to the overall financial standings of the villages in receiving attention from the district authorities, and this makes a big difference in making climate actions. The influential fishermen of the fishing villages, which are predominantly involved in capital-intensive fishing (for example, surukku valai fishers), control the "political scene" at regional levels. In contrast, the traditional small-scale fishers possess weak roles in the political parties and, therefore, they are frequently unable to negotiate with the political parties to represent their interests. Within them, fisherwomen are even more marginalized, and their functions in the parties are minor and negligibly less. As this study shows, most fisherwomen in the Coromandel coast of Tamil Nadu do not have strong political preferences over ideologies though they have preferences over political leaders and leanings. Typically, they vote in favor of the male members (mostly husbands) interests and directives (mostly husbands). Due to technological advancements and improvements in average literacy among women in the state, such trends have been in flux for the past decade.

Women's self-help groups

Women's SHGs have been active across coastal Tamil Nadu for around two decades. A few international and national NGOs and fund donors continue to support their activities for the well-being of marine fishers across the study villages. It is a semi-formal local institution, given its prominent microfinance functions (Bharamappanavara, 2013). Since the tsunami, they have been sighted in the fishing communities of the Nagapattinam district. It has increased the social status of fisherwomen and trained them to build social networks among themselves compared to previous decades. The overall socioeconomic development of fishing villages has also boosted women's standing. Any woman from a fishing village, regardless of occupation or caste, can join or lead a group. All members follow the rules and regulations of the women's SHG. Each women's self-help organization has a leader who, by consensus, selects a secretary and accountant. These members represent women's SHGs in their villages and negotiate with the International Fund for Agricultural Development (IFAD) cluster team on their behalf. In a few cases, women from SHGs joined fishermen in social and economic activities in fishing villages.

The study identifies that even small fishing villages such as Kodiyampalayam and Chinnamedu possess active and committed women's SHGs. A few foreign and national financial donors continue to support Coastal Tamil Nadu. Across the tsunami-hit fishing villages of the Coromandel coast of Tamil Nadu, the IFAD provides substantial material and non-material support to the women's SHGs. Thereby, they play important roles in offering low-interest loans to fishing households. Fisherwomen find it very useful, even though they consider these loans insufficient to meet the needs of their households, especially during rough seasons. The goal of IFAD is "to build self-reliant coastal communities, resilient to shocks and able to manage their livelihoods sustainably." It has worked with the state government of Tamil Nadu to develop various projects aimed at improving the lives of coastal villages affected by the 2004 Indian Ocean Tsunami. IFAD has efficiently implemented programs to improve the living conditions of coastal fishers. IFAD staff on the ground regularly trains fisherwomen on how to develop and organize women's SHGs. They support women's self-employment by microfinancing female fish vendors and processors (loans). The heads of women's SHGs are in charge of collecting and depositing funds from fisherwomen in the banks. The bank offers accumulated loans/credit. Women's SHGs must carry out these obligations without disputes or dropouts. In a nutshell, women's SHGs are intended to help fishing households garner money and build a credit system. The local women's SHG head regulates the loan scheme, and she ensures that the scheme runs smoothly in the fishing villages without any conflicts. The selection of local heads of women's SHGs primarily depends on the collective agreement of the members of SHGs and the discretion of the local heads of fishing villages. In addition, IFAD provides active women members with insurance training at various levels and develops practical techniques to respond to crises affecting their livelihoods.

Non-governmental organizations

Recent studies focusing on South Asia indicate that the income-generating programs, skill development training, and livelihood diversification projects of the NGOs have significantly promoted the economic and social capital of fishers (Islam & Morgan, 2012). In line with this, coastal Tamil Nadu has had similar positive results from the contributions of NGOs. Since the 2004 tsunami disaster, NGOs (both local and foreign) have begun playing various roles along Tamil Nadu's coast. The active contributions of non-governmental organizations (NGOs) have already resulted in major transformations in the socioeconomic lives of marine fishermen. It is appropriate to be termed semi-formal local institutions in light of their functions and background. Conversely, this book argues that these semi-formal local institutions – NGOs – are not grassroots organizations in most cases. This is because these NGOs' headquarters and branch offices are in taluk/district/state capitals but not in the fishing villages. The staff of the NGOs is also mostly from different regions of the district but not from the Pattinavar community. This study identifies that the credit assistance for women's SHGs enhances the ties between external agencies and the fishing community in multiple ways, and thereby, it opens up new avenues for interaction with the outside world and allows the fisherwomen to contribute financially (though it is minimal) to their households. At the same time, this study underlines that the major NGOs that support and work for the welfare of the marine fishers of coastal Tamil Nadu have not yet begun to recognize their climate vulnerability. So, these institutions focus on reducing poverty and helping people make a living, which indirectly helped fishers adapt to climate change in some ways. These semi-formal institutions are frequently afflicted by insufficient money and an absence of competent personnel. Moreover, these local institutions lack strategies and integrated plans for addressing the effects of climate change through collaboration with other local institutions in fishing villages.

Other groups

Some other small groups, such as the fan clubs of film actors, men's SHGs, and disaster management groups, are also functional in the study villages. However, the activities and resource support of these groups are random, meager, and are not systematic.

Overall insights of the local institutions

The study reveals that all the study villages lack basic infrastructure in one way or the other, limiting fishers' adaptation to climate change and risks. Some villages (Kodiyampalayam, Koozhayar, and Madavamedu) have enough physical space for disaster risk reduction training and capacity-building programs to address climate change and risks, but the village uur panchayat heads were not so interested. In the aftermath of the 2004 tsunami, as a part of tsunami recovery and rehabilitation programs, the humanitarian agencies provided a single fishing vessel (FRP-OBM unit) to four (or more, depending on the situation) fishermen from the same village, and the fishermen had begun to share the benefits and losses of fishing by working together. Several years have passed since the 2004 tsunami hit Nagapattinam. However, many of the livelihoods of small-scale fishers of these tsunami-hit fishing villages continue to rely on the fishing vessels they acquired as part of tsunami rehabilitation efforts. A few fishermen are still using life jackets they got after the tsunami. Most fishing households (95%), regardless of socioeconomic status, lack the necessary safety precautions and training to respond to imminent threats, indicating a lack of sense of disaster preparedness. Since the devastating tsunami, the fishermen of this region have been hit by major cyclones and heavy rains over the years. In 2016 or 2017, the state government reportedly provided disaster management training to the fishers of some villages (including Chinnamedu). As the key informants reported, the training was erratic and ineffective though it provided a small benefit to their fishers' populace. One contributor to this issue is the fishers: According to Pattinvar fishers, fishing is the primary focus. The fishers of financially stable villages such as Madavamedu and Kottaimedu usually show minimal enthusiasm for the state's efforts. Consequently, fishers have shown a dismal lack of disaster preparedness, with Chavadikuppam faring the least. Overall, this book identifies fisherwomen who started realizing the importance of contributing to their families after women's SHGs got involved (though this is unrelated to disaster risk reduction and climate actions). Over the past decade, women's SHGs have succeeded in mobilizing women despite a lack of resources and support from male-dominated local institutions.

Scope of the local institutions in climate adaptation: field discussions

Pattinavars feel an inextricable connection to the sea and the coast, to which they owe much of their identity and their way of life. No matter their socioeconomic standing, fishers have limited options for adapting to climate change risks and changes, regardless of their perceptions and experiences with climate change on different scales. This chapter highlights that the local institutions have strong potential and scope in making climate awareness campaigns, place-based adaptation efforts, and incorporating climate change at various stages of local planning. This will aid in maintaining the well-being of the fishers and the stability of the community in the face of climate change. Uur panchayats, fishermen cooperative societies, women's SHGs, and fisherwomen cooperative societies (although weak) are rooted in their communities. Semi-formal local institutions such as SHGs and NGOs have the ability to increase climate change knowledge among fishers because they have strong networks both within and across fishing villages. The uur panchayat leaders of Kodiyampalayam and Chinnamedu, both small fishing villages, repeatedly complained about the lack of financial stability and capital in their local institutions. The uur panchayat leaders of Madavamedu and Kottaimedu claimed that most of their village's fishing households are better off financially than those in the smaller villages surrounding them. On top of that, they claimed that the local institutions benefited from financial stability. It shows that the financial stability of fishing households is crucial for community-based climate adaptation measures. As discussed in the prior literature, shared values, social capital, and effective institutional arrangements among the local institutions fostered conflict-free resource sharing, thereby securing the fishermen's livelihoods during lean and fish ban periods (Charles, 2012; Weeratunge et al., 2013; Nayak, 2015). In Kodiyampalayam, for instance, there was less competition and greater mutual understanding among local institutions, facilitating the sharing of resources and territories. It somewhat reduces vulnerability during lean or rough seasons, thereby promoting the fishers' coping capacity. For instance, fishermen from Kodiyampalayam use catamarans (engineless) for fishing in the Kollidam River during rough seasons and fish-ban periods. It reduces the strain on their livelihoods and increases their adaptability to climate change. In the cases of Madavamedu and Kottaimedu, the marine fishermen's social ties drove them to invest heavily in surukku valai fishing techniques, resulting in increased physical and financial capital. The political influence of these villages in the taluk and district is impacted by the financial stability of their marine fishermen and local institutions (as the unified framework depicted in Chapter 2). These villages, for instance, recently obtained government and nNGO assistance to construct a coastal bio-shield in order to enhance public physical assets such as road transportation. On the contrary, road access is limited in small fishing villages, and it decreases the resiliency of the small fishing villages to climate hazards and change.

From the insights gained from tsunami-hit fishing villages, this chapter advocates that the leadership, beliefs, and shared values of the customary local institutions have substantially affected climate change adaptation initiatives, particularly coastal defense debates. Incorporating climate change into fishers' everyday conversations requires a solid information base on the climate vulnerability of their villages and their livelihoods, which is conspicuously absent in small fishing villages. Fishing villages with significant financial and physical capital and strong community leadership have had significant success in this area. Pattinavar fishermen have minimal livelihood diversification, according to Coulthard (2008). The outcomes of this study confirm that, except for marine fishing, Pattinavars have limited working knowledge and skills to work in non-fishing sectors. Aside from that, they are more interested in getting work abroad for a living. As stated earlier, the local institutions of fishers are highly receptive to hearing the problems of fishing households and would assist them by contacting local and state administrations through their customary routes. Uur panchayats of all the fishing villages hold meetings on a regular basis and make decisions based on input from local fishermen and senior fishermen. The meetings were convened for a variety of reasons; in some circumstances, urgent meetings can be called to learn about the perspectives of fishermen, such as before looming climate extreme events. Local institution heads said that they would not support diversifying livelihoods in fishing villages by expressing concern about safeguarding fishermen's integration and attachment to fishing. Nayak (2017) recommends the following measures for fishers facing a livelihood crisis in Chilika Lagoon: (i) Coping for subsistence, (ii) intensification, (iii) extensification, (iv) outmigration, and (v) diversification. This study partly agrees with prior research's recommendations for addressing the fishers' livelihood crisis (Nayak, 2017) and shows a different picture. Diversifying small-scale fishery livelihoods is an effective short-term technique for dealing with environmental shock. In contrast, this study reveals that fishers' local institutions are not in favor of diversifying their livelihoods since they strongly believe that short-term diversification tactics will separate fishers from the occupation and have long-term negative repercussions.

Faist (2018) discusses three generations of climate change and migration studies and found that the newer generation studies give the state and other collective actors the least responsibility for migrants. It views migration as the result of migrants' preventative actions. According to small-scale fishers, it is impossible to conceive of any long-term adaptation techniques or alternative livelihoods to combat the effects of climate change in the absence of external assistance. Seasonal migration is a viable option, as evidenced by several studies conducted worldwide, especially in India (Salagrama & Koriya, 2007; Badjeck, 2008; Vivekanandan, 2010; Islam & Herbeck, 2013). Backed by empirical evidence, this book contends that migration doesn't always need to indicate vulnerability. Migration can also be considered a type of adaptation (Bogardi & Warner, 2009). According to Koko Warner, migration increases people's resilience in the context of varied solutions to climate change. According to Warner (2010), there are differing perspectives on migration and whether or not it is deemed adaptive. The interviews and discussions with the marine fishermen show that they are more interested in taking jobs abroad through their social networks than diversifying livelihoods in fishing villages. The secretary of the Koozhayar village fishermen cooperative societies highlighted why fishers are increasingly interested in work chances in other nations. He was one of the important key informants for this study. For the past 25 years, he has been a fisherman and is now involved in fish trading. He said:

> Compared to our predecessors, we have the luxury of possessing televisions, mobile phones, and laptops (a few households). However, our hamlet is in a more backward state than the neighboring fishing communities that rely primarily on mechanized fishing. Surukku valai fishing methods are something we have recently dabbled in. Nonetheless, our income is minimal, and we frequently end up in debt during difficult times. In general, our Pattinavar fishermen lack a savings mindset. New investments in the fishing industry are not an option for us because the sea has already been cleaned away by mechanized fishers. We, the fishermen, are catching what they have left behind. Fishing has already evolved into a complicated and uncertain profession. We don't want our misery to be passed down to future generations. Therefore, we actively urge our children to get an education and abandon this hereditary occupation. That's why we are continuously on the lookout for work abroad. That is the most effective way to both earn and save money.

It is not up to the local fishing institutions to make a decision on whether or not returning foreign fishermen should continue fishing in their home waters.

Fishermen have complete freedom to leave and return to their local villages to continue fishing. The rule-making principle is the cornerstone of local institutions (North, 1991). Interestingly, refraining from regulating the working norms of migrant fishermen by local institutions functions as a supporting mechanism, increasing fishers' adaptive capacity in response to climate change. Notably, the key actors of fishers' local institutions stated that the population of fishing communities makes a significant difference when approaching the state and bureaucrats. Kodiyampalayam's uur panchayat head said:

> The state tends to be more responsive to the problems of fishing villages with a high population density. As you can see, the highly populated fishing villages and those that rely heavily on mechanized fishing along this coast have better roads and infrastructure than the smaller fishing villages.

The fishing villages of the Coromandel coast of South India are well-connected through kinship and family networks. Therefore, studies in the recent past argue that the stock of social capital among marine fishers is generally similar along the entire coastal stretch. Moving beyond this conventional understanding, this book argues that though the fishing villages with a low population have strong social cohesion within the fishing villages, their social networks across the coastal stretch are relatively limited and scattered than the fishing villages with substantial populations. Also, their collective investment in fishery occupation is less than that of populous fishing communities that engage in mechanized fishing. The densely populated fishing villages can capitalize on their human capital by making collective investments in mechanized fishing. In a range of ways, the small coastal fishing communities are vulnerable to climate-related risks. This study shows that these villages lack the necessary "capitals" to adapt to climate change and risks. Every year, severe and unexpected weather shocks damage expensive fishing vessels and gears of Madavamedu and Kottaimedu. Nevertheless, they are resilient enough to withstand the negative effects of climate risks and change. This is due to the responsiveness and shrewdness of their local institutions in reaching out to the state and bureaucracy in the event of severe damage. In the case of small fishing settlements, this capacity is clearly limited. As an example, political parties wield considerable power over other local institutions in Kodiyampalayam. However, their political clout has not translated into climate adaptation actions. In Madavamedu, on the other hand, political parties are restrained from meddling with the duties and activities of other local institutions. Instead, uur panchayat leaders successfully reached out to the taluk and district governments by working with political parties. In recent years, the uur panchayat of Madavamedu mobilized the local fishers to reach out to the district administration, showing that it has the potential to claim their space in district-level administration in the future, which is only conceivable for big, mechanized fishing villages.

Erosion, coastal dunes, and discussions on adaptation

In response to the devastation caused by the 2004 Indian Ocean Tsunami, the state government of Tamil Nadu began constructing seawalls and groynes to safeguard the coastlines and coastal communities. The M.S. Swaminathan Research Foundation[4] continues to work on the construction of mangrove and non-mangrove plantations, also known as coast shelterbelts. It prevents sand erosion by binding the sand. Since the 1970s, the state forest departments of India have taken measures to increase the number of shelter beds in coastal regions. In 1977, a severe cyclone struck the coastal areas of the district of Nagapattinam, causing significant damage to the district. After the incident, the Tamil Nadu Forest department began constructing shelter beds from Puthupattinam to Thirumullaivasal in the Nagapattinam district (the study villages of this research are very close to this area). Along with that, the Tamil Nadu Forest Department began growing casuarina along the state's coasts. These shelter beds and other adaptive measures played a significant role in protecting fishing hamlets during the 2004 tsunami. Following the 2004 Indian Ocean Tsunami, the World Bank-funded Emergency Tsunami Reconstruction Project also began to support the growth of mangrove plantations and shelter beds along Tamil Nadu's coasts (Selvam et al., 2005). According to Sridhar et al. (2006), the entire process of growing casuarina was conducted without the proper implementation of guidelines, and it lacked scientific rigor. They criticize and question the strong recommendations of the Swaminathan committee and note that the continuous growth of bio-shields along the coast alienates fishermen from the shoreline, as the shields could impede the availability and accessibility of the beach. They contend that the Swaminathan committee greatly overstated how much people in the area want mangroves to grow more on the beach.

Figure 6.3 Coastal bio-shields. Place: Madavamedu beach.
Picture credit: Devendraraj Madhanagopal.

Contrary to the critiques of Sridhar et al. (2006), the interviews and discussions with the local respondents show that fishers are largely in favor of mangrove bio-shields as they are increasingly threatened by coastal erosion, breaching, loss of dunes, and the combined effects of these factors. According to them, mangrove bio-shields play a crucial role in slowing down the speed of ocean waves during cyclones and natural disasters (for instance, the 2004 Indian Ocean Tsunami and the 2011 Thane cyclone). It backs up the findings of Parthasarathy and Gupta (2013) on the perceptions of coastal communities in Kachchh in Gujarat about the benefits of mangroves in reducing cyclone vulnerability. MSSRF's recent brief statement validates the above argument that mangroves acted as a natural bio-shield and protected the lives and property of fishermen from the devastation caused by the 2018 Gaja Cyclone.[5] Fisher respondents noted that mangrove bio-shields have not caused them to feel estranged from the sea and the coast. Local leaders with political affiliations echoed the sentiments of fishermen when discussing mangroves and their protective effects against climate change and extreme weather. Fishers' perspectives on mangroves essentially outline how mangroves reduce vulnerability during climate extremes and disasters. However, field interviews with the respondents suggest that they have not identified any direct economic benefits from protecting and expanding mangroves, as Viswanathan (2013) reported in his study of coastal Gujarat.

Being aware of the increasing threats posed by coastal erosion to their fishing and living spaces, the fishers repeatedly petitioned officials and state administrations to build seawalls along the coasts of their villages. The seawalls are satisfactory to the fishermen, despite criticisms in terms of sustainability and security. A respondent of Koozhayar stated that, recently (sometime in 2015 or 2016), three fishermen from a nearby fishing village died in accidents caused by seawalls, and this has instilled fear and insecurity in the psyches of fishermen throughout the region. The account of a senior uur panchayat leader of Koozhayar is as follows:

> Prior to roughly two years (sometime in 2016 or 2017), a handful of similar accidents had occurred in the villages of Thirumullaivasal and Thoduvaai. Three village fishermen were killed by seawalls. They were navigating their fishing vessels to reach the coast, and they attempted to board their vessels in the river, which is very close to the coast. A heavy sea-tide unexpectedly arose and flipped the boats, sending the fishermen crashing into the sea walls and leaving them with serious head injuries. They passed away, sadly. Since this tragedy has occurred, we have learned the hard way about the hazards posed by seawalls, particularly during stormy times of year. However, the state administration's plans to build sea walls along the coast are inevitable, and we must support them.

Aware of the dangers of coastal erosion, however, fishermen support the construction of seawalls along their coastline. Discussions with fishermen on the ground reveal that the vast majority of them have a favorable outlook on the

130 *Local institutions*

Figure 6.4 Focus group discussions with fishermen. Place: Madavamedu.
Picture credit: Clicked by a fisherman of Madavamedu.

construction of seawalls, despite knowing that this practice is unsustainable and will have negative knock-on effects on their ability to make a living. An elderly fisherman from Madavamedu shared a similar view on the importance of sea walls:

> The shoreline moves about 15 meters yearly, and it has moved a lot more since the tsunami disaster. Our ships and equipment are frequently damaged during the rough seasons. Although seawalls pose risks, they can be avoided by constructing a separate path in the nearby river for boarding our boats.

The head of the Chinnamedu fishermen's co-op society explained why building sea walls is the best and quickest way to deal with erosion in the coming years:

> Regarding sea walls and their effects, we consistently hold divergent opinions. In practice, the negative effects of sea walls can be avoided by constructing a separate boarding path in a nearby river. Chinnamedu is the only fishing village in this coastal region without seawalls. Since the state government has already built sea walls in Vanagiri and Thirumullaivasal, this village is even more at risk from erosion. In recent years, the repressed

sea waves from the neighboring villages have begun to pound our beach with greater ferocity than in the past. For example, this village was severely impacted by the 2015 massive rains because it lacks a separate channel that allows rainwater to mix with the sea. That's why it's so awful here: rainwater pools for more than four days after it falls. We drained the rainwater from our community ourselves, at our own expense. In order to prevent the destruction of our village, we are calling on the district administration to speed up the building of sea walls.

Similarly, during fieldwork, the Madavamedu fishermen expressed the need for seawalls. An experienced fisherman from Madavamedu raised the following concern:

The effects of coastal erosion over the past two decades have been dramatic. We lose about fifteen metres of coastline per year due to erosion and subsidence, a trend that has accelerated in the wake of the 2004 tsunami. It is therefore imperative that we build seawalls along our coastline. In order to accomplish this, numerous petitions were submitted through our uur panchayats to various government officials (MLAs, ministers), and we have since learned that the necessary funds have been approved by the government. However, it is currently in progress.

Figure 6.5 Focus group discussions with the senior fishermen and local leaders. Place: Chinnamedu beach.

Picture credit: Devendraraj Madhanagopal.

132 *Local institutions*

To combat erosion and protect their fishing assets, the uur panchayats and fishermen cooperative societies of Madavamedu collaborated with the local government to build coastal dunes (see Figure 6.6). Madavamedu fishers continue working together to manage the dunes for the past ten years. As recounted by Madavamedu's uur panchayat leader:

> We, the villagers, started building coastal dunes on our beach in 2006 to protect our village from the climate crisis. We obtained backing for this work from the local government in addition to the local fishing community. During the rough seasons, however, it is frequently eroded. Therefore, we have no choice but to maintain it continuously. The MGNREGA program is primarily used to keep the dunes in good shape. In such a circumstance, the government pays us the minimum wage. We have had a couple of instances where we could not get the government's backing to keep the dunes in good shape. In such cases, we must pay fishers using local monies. We benefit greatly from the dunes. It protected this village during the 2011 Thane cyclone and the most recent heavy rains in 2015.

Along with seawalls, most fishermen agree that dunes are the most practical way to deal with coastal erosion and climate change. The dunes of the coastal fishing communities were pummeled by tsunami waves. Since the tsunami, fishermen have reported a dramatic decrease in their fishing and living spaces due to coastal erosion. Prior to the 2004 tsunami, coastal dunes acted as a formidable force to prevent the sea from entering the village. The dunes were

Figure 6.6 Coastal sand dunes made by fishers. Place: Madavamedu.
Picture credit: Devendraraj Madhanagopal.

severely damaged by tsunami waves, and as a result, the surrounding villages are in danger (Sheth et al., 2006; Mascarenhas, 2006; Kumaraperumal et al., 2007; Karan & Shanmugam, 2011, pp. 16–17; Ramana Murthy et al., 2012). Since the 2004 Indian Ocean Tsunami disaster, both the state government of Tamil Nadu and fishers have come to appreciate the importance of dunes as a major defense against tsunami inundation of fishing villages. Also, it's highly integrated with fishers' lives (Namboothri et al., 2008). In a similar vein, a Kodiyampalayam uur panchayat leader voiced his concern for and appreciation of dunes. He makes the following assertion about sand dunes:

> We were fortunate to have dunes on the beach of our village, as they acted as a major barrier against the effects of the tsunami and protected our village. However, the dunes along our coast were severely damaged by the tsunami waves. The dunes on our beach have been eroding since then, largely as a result of climate change and extreme weather conditions. I am aware of the constant danger that faces our coastlines. On the outskirts of the village, there is a small seawall that is about 5 feet high and about 100 meters long, but this is insufficient to protect the village from the effects of climate events. One small bridge is the only connection between this village and the outside world. If a disaster occurs in the future, no one will be able to save us or our assets.

Overall, the fishers across the study area were unanimous in their support for sea walls to protect their coasts and fishing, echoing the stories told by fishermen and local leaders. Some fishermen wanted to see the harbor constructed. Yet they are realistic about the strength of the "capitals" at their disposal in order to bargain with the state. Many of the local leaders I met expressed concern about the negative effects of climate change; they also stated that they lack the resources to deal effectively with the risks they face. Furthermore, they are cognizant of the inadequacy and insufficiency of their "capitals" in responding to climate change. To adapt to erosion and other climate change impacts, local leaders of fishing villages have said they need help from external agencies (NGOs) and support from the state government.

When asked about the specifics of fund allocation and other sustainable community-based options for erosion prevention, an uur panchayat leader from Chavadikuppam village described:

> We, the local leaders, have no idea what the government has sanctioned to the villages to build the groynes and rubble-mound sea walls. The fund details are known to bureaucrats and MLAs who regularly visit our village to survey the coasts. Even the head of our village's panchayat does not know about the government's allocated funds and plans, and we could not care less because it is not our responsibility to find out. Our government rarely invites us to take part in debates about sea wall construction. What we really need is for the state to do some "good" things for our fishers.

Limits of local institutions in climate adaptation

Irrespective of their financial standing, all the tsunami-hit fishing villages on this coast are vulnerable to climate change and risks. The marine fishers of Chinnamedu, Kodiyampalayam, and Chavadikuppam are particularly vulnerable to coastal erosion. For the foreseeable future, coastal erosion poses a severe threat to the livelihoods of fishers across the study regions. This study, however, finds that the adaptive capacity of fishers' local institutions to climate change varies according to their economic capital and political influence. The accounts of the local leaders reveal that, regardless of their socioeconomic levels, they acknowledge that their fishing livelihoods and coastal areas have been subjected to tremendous stress in recent years. However, the strategic responses of their fishing villages have been inconsistent and inadequate due to various factors. They agreed that their inadequate "capital" limited the ability of their local institutions to respond to climate change. It accords with the previous discussions focused on different regions of the world (Nordgren et al., 2016). Pointing out this, this chapter highlights that the local institutions of fishermen in all fishing villages play passive roles in climate change communication, limiting the capacity of fishing villages to construct and maintain a network with the district administration for climate change planning.

For example, the most recent climatic stress was the extreme rains of 2015, which caused enormous financial losses to the fishers along this coastal stretch. For over three months, it negatively impacted the fishery and the livelihoods of fishers and fish traders. Fishers, regardless of socioeconomic status, rely heavily on state-sponsored schemes and welfare assistance schemes to sustain their livelihoods. The local institutions of small fishing villages had a few informal gatherings to address their livelihood concerns and attempted to contact the state government. Nonetheless, their adaptive capacity to respond to this climate risk was impeded by a lack of resources and awareness, as well as a lack of external help. In Tamil Nadu, constructing seawalls is a crucial adaptation response to coastal erosion by the state government. The government of Tamil Nadu has already built seawalls along the Coromandel Coast of Tamil Nadu in several fishing villages, and the administration is coming up with multiple plans to secure the coasts and the fishers' livelihoods in the future. In recent years, the local leaders of various study villages requested the district administration to build seawalls, and the process is still ongoing. Karlsson and Hovelsrud (2015), for example, highlight how coastal communities in Belize work together to minimize coastal erosion. They emphasized the necessity to engage with local NGOs and researchers to develop collective action campaigns to counteract erosion. Such a coordinated effort could assist coastal towns in addressing the state and accelerating the adaptation process. In contrast, in this study, the local institutions of fishers' climate adaptation measures are formal and solicitous in their strategies. They rarely built coordinated campaigns to combat erosion in the recent past.

The leaders of all the local institutions have a common understanding of the increasing complexity of climate change. They have limited resources and lobbying power with the district administration and other actors to address their climate concerns. With a few exceptions, the collective response of the leaders of local institutions, including the heads of uur panchayats, to climate change is based on formal approaches to the state government and not on informal or community-based methods. According to the findings, the leaders of fishers' local institutions throughout the research region are aware of the negative consequences of coastal erosion on their village spaces and fishery livelihoods. Due to limited networks and power, the involvement of heads of the fishers' local institutions in lobbying for seawalls has been weak. They stated that they had never been a part of the district administration committee in the planning, monitoring, and implementation of the seawall construction process. The lack of financial resources and the lack of political clout of fishers' local institutions make it difficult for small fishing villages to push the district government to expedite seawall construction on their village coasts. The study suggests that the lack of political capital of small fishing villages, in particular, reflects their inability to obtain compensation for climate-related stresses, such as the tremendous rains of 2015. In light of this, this study suggests that there is an urgent need for outside help to encourage community-wide action in response to rising sea levels, erosion, and other effects of climate change.

From 2010 to 2015, NGOs provided support, training, and credits to local institutions across the research areas to promote their fishing occupations. Local authorities solemnly agreed that any assistance coming from external actors would be suitable for fishermen in the event of climate emergency risks. All local institutions lack the resources and power to access the state and climate risks and change, said important actors in the small fishing villages. Discussions with local NGO personnel reveal that the NGOs across the study sites did not take climate change and its implications seriously. The study region has seen a lot of engagement between NGOs and local fishing institutions. It was, however, primarily limited to implementing social programs and infrastructure development projects in fishing villages. Local NGOs have in the past suggested taking action on climate change, but those suggestions were weak and did not work well enough to deal with the complex challenges of climate change and did not match with the unique circumstances of the fishing villages of this coast. Overall, the heads of local institutions, particularly small fishing villages, stated that they lack the economic and physical resources to respond to climate threats and change. Fisherwomen are underrepresented in state disaster management initiatives and are largely unaware of disaster management operations. So, they are disproportionately vulnerable to climate change. This component is consistent with previous research on other regions of the globe (Cutter, 1995; Wisner, 2003). Intriguingly, regardless of the fishing villages, the local leaders of fishers' local institutions remarked that they have limited power and resources to negotiate with the state to obtain information on erosion and the advantages of seawalls and to expedite the construction of seawalls in their village beach. To put it another way, a lack of localized data on coastal erosion's adverse effects is a significant setback for local leaders trying to combat erosion.

The government of Tamil Nadu continued to provide various welfare measures, including institutional credits, to the marine fishers during the rough season and fish-ban period (Government of Tamil Nadu, 2018, 2021, 2022a). Fishermen cooperative societies are critical in coordinating and channeling the state's institutional assistance and welfare measures to the fishers. This study emphasizes the need to tap the potential of fishermen cooperative societies and other local institutions (including women's SHGs) in responding to climate change. The state government raised the assistance amount for the fish ban period a few years ago. Respondents concurred that this credit support from the state was helpful for them to confront the crisis in their means of subsistence caused by the ban and climate risks during that period. Nonetheless, all the heads of local institutions are dissatisfied with the agreed-upon compensation figure and expressed their regret that the state's assistance is insufficient to support their families by stressing recent gross inflation. The findings indicate that the small fishing villages along the Coromandel Coast lack adequate infrastructure to promote localized disaster risk reduction activities. The primary disaster management training that the district administration gives to the fishers is clearly not enough to deal with any climate-related risks that may happen in the future. Interviews and discussions with key actors from local institutions on the first three phases of disaster preparedness (mitigation, preparedness, and response) found that the status of disaster preparedness and promoting the community resilience of fishers was significantly weak (Madhanagopal, 2018). As scholars have pointed out, the social barriers that women and children face make them disproportionately vulnerable to climate extremes (Cutter, 1995). These discussions are detailed in the following chapter (Chapter 8).

Conclusion

The process of climate change adaptation underpins both the sociocultural and political aspects of local populations and institutions. Climate change perceptions and adaptive capacity in local communities are influenced by power relations, conflicts, and the scale of interactions among local institutions (Eriksen & Lind, 2009; Wolf, 2011; Eguavoen et al., 2015). Literature highlights local institutions, community leadership, local politics, networks, people-place interactions, and traditional knowledge for successful adaptation to climate change (Amundsen, 2012). To be effective in climate change adaptation, key actors must thoroughly understand local politics, socioeconomic inequities, and political possibilities in society rather than simply reinforcing top-down adaptation strategies (Eriksen et al., 2015). Jentoft (2004) emphasizes that the human, social, and cultural underpinnings of society must be considered while studying fisheries governance systems. Experts have proposed frameworks for increasing knowledge, assets, and institutions of fishing communities to strengthen the resilience and adaptive capacity of marine fishing communities (Marshall et al., 2010; Armitage et al., 2017). In coastal Uruguay research, Nagy et al. (2015) argue that the role of local champions in climate change adaptation might be a bridge between the community and the state and private players. Agreeing with this, this book advocates it is vital to have local leaders along each fishing village on the Coromandel coast of South India who can work with state and external actors to help implement climate change adaptation strategies. Furthermore, the wide disparity in climatic knowledge, socioeconomic resources, and political power in local communities is highlighted in this research. This study indicates fishers' local institutions should spread climate awareness and spearhead climate action in fishing villages.

The financial and human capital, in addition to the political capitals of village leaders, define the opportunities and constraints for developing local adaptation practices. Aside from a few minor variations in the roles of the various customary governance systems, village governance is essentially identical. Human capital, place-based vulnerability, networks, knowledge, and information make a huge difference. As detailed in this chapter, small fishing villages lack the financial means and political clout to persuade the state government to speed up climate adaptation initiatives, making them more vulnerable to climate change. The mechanized fishing villages have sufficient financial resources to explore and execute workable solutions to the various threats to their way of life, including climate change. With such a prosperous economy, marine fishers are discussing the impacts of coastal erosion, an apparent effect of climate change, in their own backyard. This has resulted in collective actions at the local level to deal with and adapt to climate change. To provide an immediate example, the coastal sand dunes that are consistently maintained by Madavamedu's local institutions are the apparent result. In a few instances, the responsibilities and activities of the fishers' local institutions have assisted the coastal fishers' livelihoods indirectly. According to a previous study, fishers improve their adaptive capacity to withstand climate change (Brown & Sonwa, 2015). Most significant actors in the study villages bemoaned their workload and claimed they lacked time to commit to a climate change strategy. Also, they thought the government should take the lead in preparing for the effects of climate change. It is true, but only up to a point. However, in-depth interviews with key players in fishermen's local institutions revealed that these institutions concentrate more on securing credit support from the state administration than working with the state and NGOs to make fishing communities more resilient to climate change. Given the socioeconomic and governance contexts of the fishing villages, this book suggests that merely providing credit support from the state and external actors to the fishing villages is not the most effective method for fostering climate-resilient marine fishing communities. This research was focused and selective; it did not seek to provide robust empirical evidence to investigate "power conflicts" among local institutions and their immediate effects on adaptive decision-making in Tamil Nadu's fishing communities. On the other hand, this chapter discusses how recent research has concentrated chiefly on uur panchayats while ignoring other local institutions that play essential roles in adapting to climate change. In this light, there is a pressing need for more regional-level studies on the coasts of India, specifically examining the strategies of local institutions that are leading the charge in terms of local climate action. State and NGOs can help impoverished institutions coordinate their responses to disaster preparedness and climate change adaptation. This study highlights the need for multi-level coordination and subnational action in climate change planning (Jörgensen et al., 2015), with a particular focus on coastal regions, in light of research showing that local and regional governments and NGOs play minimal roles in climate change planning and policymaking in India (Beermann et al., 2016). State and external agencies working on adaptation initiatives with coastal fishing communities should be well-versed in the social, economic, and political realities that marine fishers face. However, up until now, no single scholarship has exclusively examined the localized strategies of political parties and fishers' cooperative societies in climate action in the context of coastal Tamil Nadu. This means that both of these community institutions have been under or ignored in previous studies. The book also makes the case that fishers cooperative societies are not given enough priority by the state when it comes to local disaster relief and climate change actions. The welfare initiatives of the state for fishermen are carried out through a vast network of fishers' cooperative societies. Political parties are similarly influential and well-connected among fishers and fishing villages. Hence, paying specific attention and representations to all the local institutions and considering their social networks is imperative to achieve effective and sustained adaptation to climate change.

Notes

1 Throughout the chapter, I use the Tamil term *uur panchayats* to denote the "fishermen councils" of Pattinavars as it provides a direct meaning to the customary fishermen councils. In some instances, depending on the context, I have also used the term "customary village councils," "customary local institutions," and "fishermen councils" to denote the uur panchayats.
2 Even though these societies connect millions of Indian fishermen, there is a dearth of up-to-date, in-depth literature on fishermen's cooperatives in the country. The following works provide an overview of fishermen's cooperatives' background, evolution, roles, and responsibilities (Sapovadia, 2004; Mahanayak & Panigrahi, 2021).
3 The three influential political parties in Coastal Tamil Nadu are as follows: Dravida Munnetra Kazhagam (DMK), All India Anna Dravida Munnetra Kazhagam (AIADMK), and Indian National Congress (INC). In recent years, Desiya Murpokku Dravida Kazhagam (DMDK) has started getting popular along the Coromandel coast. However, DMK and AIADMK remain more powerful and influential among the fishers among all the political parties.

Local institutions 139

4 Prof. M.S. Swaminathan, an eminent agricultural scientist, established the M.S. Swaminathan Research Foundation in 1988 as a non-profit organization. This organization continues to work on mangroves, the ecological security of coastal systems, and the livelihood security of coastal communities (both fishing and farming) in Tamil Nadu, in addition to a variety of other programs.
5 The National Disaster Management Authority's study report emphasized the devastating effects of the Gaja Cyclone on the coastal communities of Nagapattinam district (NDMA, 2019). News outlets also focused extensively on how the event evoked memories of the 2004 tsunami (NDTV, 2018a).

7 Fisherwomen and their agencies

Scope and challenges in adapting to climate change

Introduction

Over the past few decades, there has been a burgeoning growth of scholarly works on gender and gender dimensions in social institutions (Acker, 1992; Kenney, 1996; Branisa et al., 2010; Rege et al., 2013). As a result, it is well-established that climate change vulnerability and adaptation are never gender-neutral and tend to be more gendered (Goh, 2012; Moosa & Tuana, 2014; Habtezion, 2016; WMO, 2019). South Asia, on the other hand, is limitedly unexplored in terms of the functions of social relations in climate change adaptation due to inequality stemming from both cultural and gender differences in society (Onta & Resurreccion, 2011). Climate change and the increasing rise of industrial fishing are degrading the state of marine fishing grounds along the coast of India, as evidenced by accumulating studies. For example, Salagrama (2012) reports the many vulnerabilities of fisherwomen in the face of climate change based on empirical evidence from four Indian coastal states. Pearse (2017) argues that a lack of understanding of women and gender makes it challenging to assess women's agency in climate action. Recent extensive research (Rao et al., 2019) examines how women's agency and social links influence climate change adaptation responses in various climate hotspots around the world. Nonetheless, the primary focus of this study is on rural women who depend on rural-based livelihoods. Thus, in global fisheries research, there has been little attention on understanding the many elements of gender systems and gendered globalization. In the past, publishing articles on women and gender issues in the fishing industry in journals with a high impact was challenging. Such trends have already changed, and scientific groups have begun documenting the gendered experiences of fisheries and coastal communities in many geographic zones (Hapke, 2012; Frangoudes & Gerrard, 2018). Despite this, popular masculinist discourses have effectively excluded women from leadership positions in climate change adaptation. Gender identities and inequities are deeply ingrained in Indian families, castes, classes, and social status (Geertz, 2004; Thomas, 2004; Raman, 2005). Gender roles in fishing, in particular, have strong sociological and historical roots. In addition to selling fish, Indian fisherwomen are required to bear children and handle all elements of domestic and social

DOI: 10.4324/9781003187691-7

reproduction. The fishing industry has long been seen as male-dominated, although women played a crucial role in the fishing industry throughout history (Thompson, 1985; Ram, 1988, p. 251; Sundar, 2010). With a few significant anthropological exceptions, little is known about the role of fisherwomen in Tamil Nadu's local governance systems and their impact on post-tsunami rehabilitation initiatives and climate change adaptation strategies. A dearth of knowledge gaps exists on the shifting gender dimensions and their intersections with local fisher institutions in the context of climate change adaptation. Focusing on the gender dimensions of fishers and how they relate to caste and class would give us a better understanding of the vulnerabilities of the fishing communities to climate change and how they adapt to it (Colwell, 2016, p. 105). This chapter explores the emergence, potentials, and limitations of fisherwomen's agencies in responding to climate change, which has not been covered in previous studies on South India's Coromandel coast.

Women's agency and empowerment

Empowerment refers to the growth of people's ability to make strategic life decisions in an environment where this ability was previously denied. To make strategic life decisions, there are three interconnected dimensions to consider: (i) Resources, (ii) agency, and (iii) achievements (Kabeer, 1999). According to Alsop et al. (2006), empowerment refers to the group/individual's capacity to make effective choices and then transform the choices into desired outcomes. This capacity to make judicious decisions is governed by two interconnected variables, namely (i) agency and (ii) opportunity structure. Agency is the ability of individuals or groups to make deliberate decisions. Conversely, the formal and informal setting in which people or an organization's function might be described as the opportunity structure. Groups are impacted by their opportunity structures, which include formal and informal institutions or game rules. The formal and informal regulations (customary norms) govern the accessibility of assets to individuals or groups and define the feasibility of changing assets into desired outcomes. As a result, even if individuals or groups can choose, they may not be able to employ agency successfully. Individuals'/groups' capacity to transform actions into desired results is influenced by the opportunity structure and institutional context. In addition, the assets and talents of individuals and groups influence the ability to make well-informed decisions. Therefore, "agency" and "empowerment" cannot be used synonymously. There are different levels of empowerment along the lines of operating structure (Alsop & Heinsohn, 2005, Alsop et al., 2006).

Women's agency and empowerment have been the subject of an increasing number of studies in recent decades (Agarwal, 2002; Kabeer, 2008; Charrad, 2010; Hanmer & Klugman, 2016). Sen (2000) notes that women are no longer passive recipients of welfare assistance from the state and external donors – women are the active "change" agents and dynamic proponents of societal transformations in their own lives as well as others (including women, men, and children). By stressing land ownership, access to capital, and the availability of resources, he claims that women's participation in economic activities may result in the expansion of their agency and empowerment. By reviewing the definitions and concepts of women's empowerment, Mishra and Tripathi (2011) note that the following terms define women's empowerment: options, choice, control, power, decision-making ability over life and over resources, freedom to make life choices, and the ability to affect one's own well-being. Based on research in Asia, Rao et al. (2017) explain the problems of connecting women's empowerment and gender equality with their assets and access to land. Gender relations, as well as the many overlapping disadvantages that men and women endure, are not equal and inflexible across geography and time. On the other hand, gender relations and identities are dynamic and diverse, acting within particular ecological and social contexts. So, to understand how gender equality works in society, it is not enough to look at how women and men divide assets and how easy it is for them to get resources. Scholarships on gender and gender aspects in social institutions (Acker, 1992; Kenney, 1996; Branisa et al., 2010; Branisa et al., 2013), as well as the intersections between gender and caste in India, have also increased in recent decades (Rege et al., 2013). Lorber (1994) argues that gender is a social construct that, unlike sex, has nothing to do with genitals and reproductive organs. Men have always been in charge of institutions and have run the vast majority of them throughout history. The only institution that gives women the space they need is the family. Examining institutions through the prism of gender might reveal how they accommodated women and how they have been shaped by and through gender (Acker, 1992). Hapke (2012) emphasizes that gender issues, gender systems, gender interactions, and gendered aspects of globalization receive less attention in the fisheries sector. Keeping this in mind, this chapter begins by discussing the gender issues of Pattinavar fisherwomen.

Gender systems of the maritime fishing communities of Tamil Nadu

Globalization, capitalism, and the 2004 Indian Ocean Tsunami all had multiple effects on the gender and social relations of fishers in South India (Hines, 2007; Gandhi, 2011; Hapke, 2012). Significant scholarly works are available to understand the gender systems and gender divisions of labor among the Mukkuvar fishers of Southern Tamil Nadu (Ram, 1992; Busby, 1995; Hapke, 1996). Aparna Sundar's work on South Indian fisheries in the Kanya Kumari district sheds light on Mukkuvars and Paravars gender systems. She discusses the changing gender roles in fishing communities in light of the emergence of microcredit and self-help groups (Sundar, 2010). The Kanyakumari district and the southern region of Kerala serve as the primary settings for these scholarly works. Thus, we can draw parallels between the gender systems of the Pattinavar fishing community and academic works investigating gender systems and gender divisions of labor in other South Indian fishing communities. Several works, such as MacDonald (2005), Pincha (2008), and Juran

(2012), look at how the 2004 Indian Ocean Tsunami affected men and women, with a focus on Tamil Nadu. There has been some discussion of gender issues in policymaking and implementation in small-scale fishing governance in recent large-scale studies. It focuses on how overfishing, falling fish stocks, and natural disasters have made fisherwomen of South India more vulnerable (for example, FAO, 2017). For centuries, the fishing villages of the Nagai-Karaikal coast have maintained strict caste- and gender-based divisions of labor. Colwell (2016) classifies the gender divisions of labor in the Pattinavar fishing communities, noting that fishermen do the bulk of the work in the ocean, including keeping everything running smoothly, catching fish, and getting them to market. Caregiving, household economics, and, for some fishing families, selling fish as head loaders are the typical limits placed on women's labor. Thus, fisherwomen participate in post-harvest activities such as drying and curing fish (Kruks-Wisner, 2011) and domestic chores. Colwell (2016) classifies gender divisions of labor in Pattinavar fishing communities, noting that men dominate fish capture and harvest while women dominate post-harvest fishing. With a particular focus on Tamil Nadu, select works examine the 2004 Indian Ocean Tsunami disaster from gender perspectives (MacDonald, 2005; Pincha, 2008; Juran, 2012). Therefore, gender issues in policymaking and implementation in small-scale fisheries governance have recently received growing attention among scholars. The vulnerability of fisherwomen in South India is discussed, emphasizing how overfishing, falling fish stocks, and natural disasters have exacerbated the problem (FAO, 2017). Significant scholarly works discuss Pattinavar women's fishery roles (Bavinck, 2001b; Gomathy, 2011; Colwell, 2016; Bavinck & Vivekanandan, 2017).

In the study on the Mukkuvar community, Ram (1988) discusses how the sexual division of labor forces fisherwomen to take on the responsibilities of domestic activities and the social production of their households. This holds for Pattinavar fishers as well. The influence of fisherwomen on the social reproduction of their fishing households (as evidenced by their participation in marriage, kinship, social, and religious practices) is sporadic and inconsistent and varies along socioeconomic lines. In addition, as identified in the fieldwork, interference by Pattinavar fishermen in the social reproduction activities of fisherwomen has also been pervasive and not uncommon. This study confirms that this is exceptionally high in fishing villages that are economically sound and stable. In recent years, the balance of gender-based labor divisions (see Table 7.1) has been in a transitional phase because of shifting livelihood patterns and rising migration levels. On the one hand, the 2004 Indian Ocean Tsunami disaster had already brought numerous positive transformations to the lives of Pattinavar fishers. On the other hand, the Pattinavar fishers continue to adhere to their gender roles, norms, rules, customs. and ethics, sometimes problematic in providing space to women in the public domain. As pointed out in the last chapter (Chapter 6), the women of small fishing villages are more interested in livelihood/income-earning opportunities, such as raising goats and participating as wage labor in the national rural employment scheme, MGNREGA.

Table 7.1 Gendered divisions of small-scale fisheries – focusing on the Pattinavar fishing community

Fishing and allied activities/ category	Women	Men
Fish capturing	✗	✓
Boat buying, maintenance, and repair	✗	✓
Fish drying, curing, and vending	✓	✗
Fish transport	✗	✓
Fish auctioning	✓	✓
Fish marketing	✓	✓
Ice production, supply, and preserving the fish	✓	✓
Supply of nets/gears and net weaving	✗	✓
Nature of the work	Less risky	Risky – It depends on the nature of fishing vessels, gears, distance, and access to information
Time of the fishing activities	Mostly day-time	Both day and night

Source: from Kruks-Wisner, 2011; Colwell, 2016; Koralagama et al., 2017; Fieldwork of the author.

Gendered institutions – post-tsunami and aftermath

In her research on the Nagapattinam district, Gandhi (2011) demonstrates how the gender norms of the Pattinavar community exacerbated the numerous vulnerabilities of women in the aftermath of the tsunami. As previously discussed in other chapters of this book, in the post-tsunami period, the influx of external actors and institutions into Tamil Nadu and Puducherry's coastal villages brought about long-lasting changes in the socioeconomic status of marine fishers. Many domestic and foreign organizations have begun funding women's self-help groups in the fishing communities along the Coromandel Coast of Tamil Nadu and Puducherry. It has been effective at increasing women's participation in the domestic affairs of fishing villages. In addition, the last two decades have seen revolutionary shifts in the way of life for the coastal fishing communities across all of South India due to the proliferation of information technology and the dissemination of new knowledge (Fihl, 2014). However, interviews and conversations with fisherwomen during fieldwork suggest that women's engagement in local politics and decision-making processes is largely unchanged from the past and remains stagnant. Some of the women I talked to had been secretaries of a fisherwomen cooperative society in the past. They said that local decisions about fisheries management, disaster risk reduction, and building coastal walls had been made by male leaders and senior fishermen without any input from fisherwomen.

Women reservation in gram panchayats (statutory)

In India, the 73rd and 74th constitutional amendments grant legal status to the gram panchayats and empower the elected members of the local panchayats to make decisions at the grassroots level and on activities that directly affect the lives of the people (Government of India, 2007). It introduced special provisions for the weaker sections of society (including SCs, STs, and women) and mandated that at least one-third of the seats in panchayat raj institutions must be occupied by women (Palanithurai & Ragupathy, 2008). Several authors (Hust, 2002; Thomas, 2004; Rai, 2013) discuss the political representation of women in the local self-government of India. The widely accepted positive impact of reservation quotas is its potential to lead to women's empowerment, while the principal critique is that it may force women to act as proxies for powerful males in their families and communities (Hust, 2002). Several empirical studies on decentralization stress the importance of women's participation in panchayat raj institutions (Bryld, 2001; Kudva & Misra, 2008). However, the power dynamics in local governance structures may vary by region and community (Palanithurai, 2008). Although the 73rd and 74th amendments to the Indian constitution gave statutory gram panchayats legal status, the power status of community (traditional) panchayats has remained unchallenged in many regions (Lele, 2001). While examining the factors that hinder women's participation in local village self-government, Thomas (2004) argues that "quotas" for women in local statutory institutions are necessary and successful as scholars identified (Thomas, 2004; Raabe et al., 2009), the caste, family pressure, party politics, and other socioeconomic and institutional factors can obstruct the active participation of women in formal local governance systems. The yield of the reservation to women in panchayats in India has not been successful compared to caste-based reservation quotas. Caste and gender quotas, on the other hand, have had a lot of sound effects and are a key part of empowering marginalized groups (Thomas, 2004; Geertz, 2004; Palanithurai, 2008). Thomas (2004) argues caste quotas have primarily been more successful than gender quotas for two main reasons: caste politics and the swift mobilization of caste by the political parties. Furthermore, in Indian contexts, "gender" is not accorded the same significance as caste identities, both in society and in politics. Hence, it is not surprising that the gender quota has not yielded the potential results and fundamental changes in society that the caste quota did. Representation of women and their visible presence in sociopolitical institutions could potentially enhance the gender consciousness of society. Nonetheless, it should also be noted that there is no guarantee that the mere presence of women in the sociopolitical institutions ensures their participation in the local institutions and empowers them.

Are customary village councils undermining local democracy and women's agencies?

Mandelbaum (1970), in his seminal work on Indian society and village panchayats, defines panchayats as "a means of gathering." He explains that panchayats are not a form of a formal political institution with delineated roles and responsibilities. It can be understood as the pattern of "dialectic, decision, and action." While discussing the roles of the panchayats and panchayat leaders in village society, Mandelbaum states that the panchayat leaders are expected to represent their lineage and to "act as a link between their group and the larger social echelons." The major roles of the panchayat leaders are to protect and preserve the traditional customs of their jati, society, and the sentiments of the groups they belong to, but not to alter them. In her empirical work on customary village councils in rural Karnataka, Ananth Pur (2007a) expresses a similar thought. She writes that the underlying roots of customary village panchayats lie in "traditions and customs." She categorizes two types of customary panchayats in Indian villages: (i) Panchayats in single-caste villages and (ii) panchayats in multi-caste villages. In the case of single-caste villages, the relationship among the villagers is based on the lines of lineage and kinship. Hence, customary panchayats are nothing but caste councils. In the case of multi-caste villages, the representations have been given to various caste groups, and hence, they could receive their due representations. There is a widespread belief among the urban public and outsiders that the traditional panchayats in India are archaic, undermining local democracy, inherently hierarchical, exploitative, and not representative of the people (Ananthpur, 2007b; Ananth Pur & Moore, 2010). Village councils play significant roles (prime roles, in the case of fishing villages) in dealing with the state by advocating for the well-being of local folk. Hence, in the Indian villages, the relevance of traditional panchayats continued to survive even after the widespread impacts of modernization and privatization. It should be noted that several notable transformations have happened in the roles, functions, and characteristics of rural society ever since constitutionalizing the PRIs. However, in various regions of India, the conflicts between traditional panchayats and newly formed statutory panchayats are highly visible and common (Lele, 2001; Gayathri Devi, 2006; Palanithurai, 2008; Singh & Goswami, 2010). Villagers' organizations (including customary institutions) and their local leadership could provide support to enhance social capital, which, in turn, helps the rural community to foster their democratic participation and development (Krishna, 2002). This is true to a partial extent in the fishing villages of South India. However, the questions are how influential local institutions in fishing villages deal with gender concerns; how they intersect with local institutions that are solely for women; and the implications of gender concerns in coping and adapting to climate change.

Women's position in the local institutions – reflections from the tsunami-hit fishing villages of South India

In recent decades, there has been an increase in scholarship discussing how gender differences make people more vulnerable to climate change and coastal disasters (Denton, 2002; Enarson & Fordham, 2001; Ariyabandu & Wickramasinghe, 2004; Pincha, 2008; Enarson & Chakrabarti, 2009;

Venkatasubramanian & Ramnarain, 2018). It emphasizes the importance of women's participation and efforts in building disaster and resilient climate communities. Williams (2016) proposes an activist campaign to raise and promote awareness of gender-related concerns by highlighting gender equality and equity in the overall socioeconomic development of fishing communities. Fishing regulations and policies often fail to account for or actively discriminate against women. In some ways, the findings of this study corroborate those of others that have reached similar conclusions. As the field observations demonstrate, the gender equality concerns of the fishing villages are not addressed by local groups in any contexts outside of fishing, cultural and religious activities, and domestic matters. In May of 2015, for instance, I went to a meeting of the uur panchayat in Kodiyampalayam to hear the fishermen's perspectives firsthand. Despite their economic disparities, the village males gathered to plan for the upcoming temple festival and had a heated discussion about the expenses and revenues. Older fishermen and uur panchayat representatives talked with the young fishermen, which helped settle things down. On the same day, I spoke with a few female fishers from the same hamlet about the possibility of their attending similar community gatherings. None of the women I spoke with were enthusiastic about being involved with uur panchayats.

In particular, fishermen have sole control over the formation, operation, and management of uur panchayats. All the local governing structures in the fishing communities are run without any input or participation from the women. Due to centuries of male dominance of fisheries and coastal systems, fishing in Pattinavar has its own gendered norms, practices, and sets of institutions with their own ways of styles. As a result, uur panchayats have never felt compelled to include women in the internal governance systems of fishing villages and coasts. Given the historical context, it is understandable why only fishermen dominate the internal governance systems of the villages across this coastal stretch. Therefore, the conclusion that Pattinavar norms and institutions are authoritarian and oppress women, as Bavinck (2006) notes, is a hasty one. This book also concurs with Bavinck that customary local institutions of the tsunami-hit fishing villages of South India are not wholly repressive and authoritarian except in a few rare cases.[1] Nonetheless, by demonstrating field observations, this book also highlights that the exclusion of women from local institutions continues to have various setbacks in the overall development of the fishing villages (as reported by Agarwal, 2000, in the study of community forestry organizations), making women more susceptible to climate change and its associated risks. Excluding women from domestic affairs and internal governance systems is customary and a part of the Pattinavars' practice that they have followed for generations. However, in recent years, as the fieldwork of this study documents, the local fisherwomen of tsunami-hit villages have started viewing such customary norms with dissatisfaction and frustration, though their efforts to confront them are feeble. Pattinavar societies and their uur panchayats provide opportunities for women's voices to be heard indirectly, such as through male family leaders. As noted in the previous research, in certain instances, the women communicated their difficulties and expectations to the uur panchayats via the male heads of their families (Bavinck & Vivekanandan, 2017). In some instances, however, the women of fishing villages make their voices heard through the local women's leader (who is typically an active member of women's self-help groups).

Empirical research on customary village councils in rural Karnataka (Ananth Pur, 2007a) reveals that uur panchayat leaders coexist with, interact with, and influence (informal) local women leaders in ways that are not always negative. In the same vein, this book examines the influence of uur panchayats, which may have both positive and harmful outcomes, and gives mixed findings. The discussions with the local leaders reveal that uur panchayats dominate and exemplify patriarchal social norms – both directly and indirectly. However, they are accountable and responsible for the fishing communities they represent. In some cases, however, fisherwomen complained about the uur panchayat leaders, saying they were not receptive to their concerns and that they had impeded the work of women's self-help groups, making it more difficult for those social groups to function. Due to the economic disadvantages of the fishing communities, women's active engagement in and contributions to marine fishing are essential. In small fishing villages like Chavadikuppam, Kodiyampalayam, Koozhayar, and Chinnamedu, female-headed households rely solely on earnings from head loading, MGNREGA, and women's self-help groups. Since starting to do surukku valai fishing, most fishing households in Madavamedu and Kottaimedu have seen their finances improve.

In some selected cases, the fisherwomen travel long distances as groups to the nearest fishing harbors to purchase the fish. For example, the women of Koozhaiyar travel to the nearest fish landing center, Thirumullaivasal; the women of Chinnamedu travel to Vanagiri; and the women of Chinnakottaimedu travel to the nearest fishing harbor, Pazhaiyar. All this travel has been exhausting and time-consuming for fisherwomen, and they rarely find time and interest to engage in the domestic affairs of their fishing villages. On a typical fishing day, fisherwomen purchase fish from the fishermen and traders by auction and carry them (as a headload) to the nearby towns for door-to-door sale, focusing on particular public spaces like bus stands and regular customers. The discussions with the fisherwomen respondents reveal that most male heads of the uur panchayats of study villages are apathetic toward ensuring the rights of fisherwomen. In a few cases, the rules and dictations of uur panchayats hampered the smooth functioning of their livelihoods. Some women of Chinnamedu and Madavademu have shared their discontent with the excessive interference of men over their petty occupations and headload trading. Given the exhaustive physical labor they spend on selling the fish and performing social reproduction responsibilities, their everyday activities are largely confined to their homes and nearby neighborhoods. Besides, the regular interference of fishermen and male-centric local institutions over their occupation and credit economy remains a visible obstacle to affording their time, efforts, or concerns in the domestic affairs and the internal governance systems of the fishing villages. In substantial instances, women pointed out that they rarely could afford the time and effort to engage themselves with women's self-help groups. As observed by Ram (1988, p. 253), women who are not involved in headload fish trading are younger women with more social reproduction responsibilities. Similar results are reflected in another study (Salagrama, 2006, p. 60) that examined the poverty of the coastal fishing communities of Odisha, India. Senior female respondents repeatedly stated that autonomy and external support with generous inflows of funds are practical means to ensure women's participation in responding to livelihood stressors and instilling an awareness of local participation. Some senior women lamented the exclusive domination of uur panchayat heads and the limited support from the external agencies, which is one of the major obstacles to promoting women's agencies.

Exclusion of women and its implications on adaptation to climate change

On behalf of their fishing villages, uur panchayat leaders and senior fishermen engage with the state and other institutions for social welfare benefits and livelihoods issues, such as disaster risk reduction and climate actions. When external actors are involved, women are always left out of negotiations. Excluding fisherwomen from discussions on coastal erosion and the construction of groynes and rubble-mound seawalls on the beach rendered them oblivious to existing and emerging stresses to their fishing livelihoods, making them unprepared to respond to the stressors. In many cases, women are uninformed of negotiations and events involving fishermen and uur panchayats. However, this study finds no evidence to support the claim that these women are targeted and brutally excluded from the participation and decision-making processes of local institutions. Except in rare situations, women of all socioeconomic classes are unwilling to join uur panchayats. As expressed by a senior woman in Kodiyampalayam,

> In general, our Pattinavar women are not interested in understanding and engaging themselves with local events because of their widespread ignorance.

Given changing weather patterns and vulnerability to climate change and disasters, it's crucial to understand how excluding fisherwomen from public spaces and apathy from local institutions have rendered the fishing community more vulnerable to climate change. In most cases, the uur panchayats intervene in the activities of external organizations meant for women, even if they are for a good purpose and are non-political. Therefore, it is uncommon for the NGOs to facilitate meetings for the local women leaders to promote women's self-help groups without prior permission from the panchayat heads. In contrast, a rare incident occurred in Chinnamedu, where the uur panchayat heads and senior fishermen welcomed the roles of NGOs in their fishing village. This was due to two main reasons: (i) Compared to their counterparts, Chinnamedu local fishers' means of subsistence and fishing grounds are under extreme strain, and (ii) women are left with no choice but to look for ways to earn income to maintain their households. Such escalating challenges to fishing livelihoods have rendered the customary institutional systems of Chinnamedu sufficiently adaptable to allow outside entities to encourage the participation of their women in local affairs. Nonetheless, the uur panchayats are cautious enough to avoid authority overlap among the institutions. In the words of an ex-active member and head of women's self-help groups in Kodiyampalayam,

> Our uur panchayat leaders are mostly non-cooperative when it comes to supporting the women to initiate any new things/projects in the village. It is the common and collective mentality of our men. It does not matter who elects as the uur panchayat leader.

This picture is much more evident in the fishing villages that are largely engaged in surukku valai fishing. For example, when I interviewed the Madavamedu fishers in 2017, the elected leader of the local self-government (local body) was a woman – I assumed the constituency was reserved for women. When I asked the villagers where the local self-government leader was, everyone pointed out that her (women's) husband is the acting leader of the panchayat constituency despite their socioeconomic and political differences, indicating that the fishers believed women were unable to deal with the internal governance systems of the fishing villages and concerns linked to fisheries management. The following is the statement of that village leader:

> I became the village president from the support of my husband. At the local level, he is the well-known person in the political party (she named it). Apart from this village, no one knows about me and what is the need to know about me since he takes care of all the responsibilities? He is the person who is aware of the needs of our fishing community. Further, for me, how does it possible to interact with the uur panchayat heads of the nearby villages?

This is the typical widespread opinion shared by the fisherwomen who are in better economic conditions. Discussions with the uur panchayat heads suggest that they are not interested in sharing power with the women. After being asked why he felt that the participation of women in the local institutions is unnecessary, one of the uur panchayat leaders of Madavamedu replied the following:

> It is our tradition that women should perform their roles in households, and we must practice these customs for generations. What is the need to separately include women in urr panchayat as they are enough to take

care of all the fishing and local activities? Women are not oppressed here. They are entirely free to do all the tasks in the fishing villages. Many of us (fishing households) provide good education to all our girl children in English-medium private schools. We have not felt the necessity to include women's participation in our panchayats.

Many uur panchayat leaders expressed similar patriarchal thoughts. The above narrative suggests that the uur panchayats recognize the need for providing good education to the next generations irrespective of gender differences. However, they are not ready to share equal rights and equal space in the decision-making structures of local governance systems. Overall, as the fieldwork shows, the fishermen are not in favor of allowing women to enter the public domain. Here, we can recall the similar observations reported by Panda (2006) in the study on the collective action of women in water management in Gujarat) for several reasons – mainly to preserve their patriarchal norms and village autonomy.

Women self-help groups in Tamil Nadu

In India, SHGs were established in the 1980s to combat poverty and empower women. NABARD (National Bank for Agriculture and Rural Development), RBI (Reserve Bank of India), and several prominent non-governmental organizations (NGOs) played critical roles in implementing SHG goals at the grassroots level. Since then, SHGs have proliferated across India with the support of the state, leading NGOs (including IFAD), banks, and cooperative agencies' support (Fernandez, 2006, p. 8). The growing interest of state and local governments, the high resilience of self-help groups, the increasing investment of banks and state governments, and the growing emphasis on microfinance systems are some of the major factors driving self-help group growth. Like this, a few factors limit the potential of self-help groups. Among them are: (i) The reluctance of banks to support this program wholeheartedly and the weak infrastructure to ensure program services and (ii) the interference of politicians in women's self-help groups, as they have begun to use this program to cultivate their "political capital." This program's primary goal is the "empowerment of women," but it seems to be shifting its attention instead to providing microcredits (Fernandez, 2006). Self-help groups were introduced as "*Magalir Thittam*" in Tamil Nadu (In Tamil: to all women). Since the tsunami, countless new self-help groups have sprouted up in the district's fishing villages (Sagayaraj, 2013). According to the National Commission of Women, women's self-help groups work toward many goals beyond microfinancing, such as educating women about government bureaucracy and giving them a forum to air their opinions and problems (NCW, n.d.). In reality, the women self-help groups in India are largely centered on providing microcredits[2] to women irrespective of their donors (Agarwal, 2001). Given this context, this book defines self-help groups as a network of women working together to empower their families, neighborhoods, and communities. These diverse networks of women of various ages, religions, regions, castes, classes, and political affiliations are supervised and administered by the state government or NGOs or both for the social, human, and economic development and well-being of their families, neighborhood, and community (definition is modified from Sagayaraj, 2013). The regions where this study was conducted are homogeneous single-caste fishing villages; consequently, most women in self-help groups are Pattinavar (with some exemptions such as Kodiyampalayam and Koozhayar). Compared to other heterogeneous multi-caste regions, the women's self-help groups in the area under study are highly networked by caste and kinship ties. Recent studies examine the role of self-help groups in Tamil Nadu's post-tsunami reconstruction efforts (Regnier, 2007; Pincha, 2008; Gandhi, 2011; Larson et al., 2015). In addition, as Agarwal (2000) argues, group homogeneity and fewer divisions among women facilitate greater cooperation in collective action than among men. Given the effects of climate change that are already happening and those that are expected to happen in the future, it is becoming more important to look into how women's self-help groups could help fishers adapt.

Credit networks of women self-help groups

The semi-formal local institutions that are exclusively meant for women in the research region are NGO-formed women self-help groups. External agencies (like IFAD) have largely operated the groups by providing training, financial support, and monitoring. In the fishing villages of Nagapattinam district, women's self-help groups have been active since the 2004 Indian Ocean Tsunami. Even before the tsunami, women's self-help groups were present but sporadic, inactive, and not networked. The results indicate that, in the aftermath of the tsunami, the number of women's self-help groups has multiplied over time. Nonetheless, SHGs continue to prioritize microcredit primarily centered on loan distribution and recovery. As pointed out by a grass widow in Chavadikuppam:

> In case of a poor fish catch, I rely on relatives and neighbours to feed my school-aged son. I find women's self-help groups and credit support useful. During lean seasons and fish prohibitions, we (Chavadikuppam) fishers receive little credit support from self-help groups. Our women's self-help groups just exist for name sake and are tragically unable to address our expanding requirements.

Fisherwomen expressed similar views in Chavadikuppam, Chinnamedu, and Chinnakoddaimedu. It indicates that the growing interest of fisherwomen in women's self-help groups is to get loans to support their families. It is not necessarily the case that the growing interest among women in acquiring loans directly reflects their participation in public actions and that this could improve the overall well-being of their fishing hamlets. From women's self-help group representatives and active members, I learned that NGOs made significant social changes along the coast over the past two decades. Along with credit support, local NGOs helped fisherwomen approach the government and bureaucrats during and after crises like floods and natural disasters. In some cases, NGOs encouraged fisherwomen to participate in discussions about seawalls and disaster risk management. For instance, the secretary of a women's self-help group in Chinnamedu reported that a local NGO encouraged fisherwomen to participate in protests against the construction of sea walls on the grounds that it would disrupt their means of subsistence. Kruks-Wisner (2011) reports similar positive effects of women's self-help groups in two Pattinavar fishing villages in Nagapattinam district, Tamil Nadu. Such collective action by women throughout the study region is contingent on a number of crucial conditions in addition to the facilitation of external entities (NGOs). It comprises adaptable local institutional arrangements, support from male heads, and social capital.

Women self-help groups and microfinance: prospects and a few questions

Sundar (2010, pp. 246–256), focusing on the Kanyakumari district, documents the growth of women's self-help groups in coastal Tamil Nadu since the 1990s. Her analysis of women's self-help groups is particularly pertinent to understanding the emergence of women's SHGs in coastal Tamil Nadu. Using an IFAD report, she explains how women's self-help groups have significantly impacted women's participation in decision-making processes and how it has boosted their confidence in their households and community over time. She also cautions that claims of increased female participation in the fishing community because of self-help groups are unclear and mixed (Sundar, 2010). Similarly, Bina Agarwal, a scholar of gender and related environmental issues, raises the issue of empowering women and closing gender gaps in society through microcredits to women's self-help groups. Land distribution and women's access to land could be more effective than microcredits in ensuring the well-being of women and their bargaining power in the community. Without questioning the existing inequality of sharing assets (land), the provision of microcredits to women may increase society's gender inequalities (Agarwal, 2001). Similarly, Ray-Bennett (2010), examines the role of microcredit in reducing women's vulnerability to disasters and highlights that the activities of SHGs and NGOs are primarily focused on providing micro debts to women, which may force them to live with debts. Hence, microcredits are not always beneficial. It could further increase the burden of an already vulnerable community due to social and gender hierarchies.

Women's agency and uur panchayats – a narrative

From 2008–2014, a local NGO (S),[3] used to provide training and aids to the SHGs of Madavamedu to grow prawns. Madavamedu fisherwomen grew prawns for a meager income. Such revenue helped fishing households cope during rough seasons, fish-ban periods, and less fish catch. Initially, it went well, but conflicts between uur panchayats and the NGO's local employees forced it to stop working in Madavamedu. A representative of women self-help group of Madavamedu elaborated the picture[4]:

> After the tsunami disaster, many non-governmental organizations (NGOs) and aid groups came to our village to provide relief and begin rebuilding. During this time, a local NGO (S) began aiding our households, focusing on fisherwomen. The uur panchayat heads and the NGO had a few disagreements over constructing free-tsunami houses. Uur panchayat believed NGO workers had swindled tsunami aid, so they denied NGO's role in our village. It is not just that particular NGO. Without the permission of the uur panchayat, no external entity is permitted to enter the village. They deny the activities of any external agency in our village if they suspect their authority is being questioned. IFAD and the local microfinance organization currently provide loans to us. Such loans are too low for self-employment. Women are interested in self-help groups and want to contribute to their families. Due to the strict regulations of uur panchayats, no progress has been made to promote the aspirations of women. We have no alternative but to rely on our uur panchayats.

The secretary of the women self-help groups of Kodiyampalayam, who has been active in organizing the groups for 20 years, has similar thoughts on uur panchayats.

Women's agency in adapting to climate change

Women self-help groups bring alternative sources of credit and new livelihood strategies to women. Equally important to the male fishing economy is the female credit economy of fisherwomen. However, it is invisible and little explored. As noted by Ram (1988, pp. 253–254), the female orientation of this parallel economy has a gendered face. It serves as a classic example of community-based strategies in the Tamil Nadu marine fishing society. Whereas the women respondents repeatedly stressed that their credit networks are insufficient to meet their livelihoods demands when there is little, or no fish caught. In multiple cases, these credit networks assist households headed by women in paying their debts. Hence, it has been of value to fisherwomen and aided the fishers during crises. All the women respondents who are active in women's self-help groups emphasized that their livelihoods and income are seriously jeopardized due to climate risks and change. However, as a senior head of a women's self-help group in Kodiyampalayam explained, around 90% of women across their village are a part of women's self-help groups. However, in the guiding strategies of women-based local institutions, climate change is not considered, whole or in part.

Collective actions of women self-help groups against seawalls

Contrary to the preceding impression, women's self-help group leaders in small fishing villages, declared that their uur panchayats support their operations. Discussions with the heads of self-help groups in Chinnamedu and Chavadikuppam revealed that regardless of the type of external agency, the uur panchayats and the fishermen cooperative society are receptive to them and offer their cooperation to carry out productive activities in their villages. For instance, women self-help groups of "S"[5] and "I" were being encouraged by the uur panchayat heads of Chinnamedu. The secretary of Chinnamedu's women's self-help groups interacted with the external agency and collectively opposed the construction of sea walls, claiming that it would severely disrupt fishermen's fishing activities. Even though the NGO mobilized women to protest seawalls, fishermen and uur panchayats did not act against them. In contrast, they joined with them to support the cause. In contrast, the financially stable women of Madavamedu were unaware of the sea walls that have been a hot topic among fishers along the coast. As fishermen in small fishing villages like Chinnamedu and Koozhayar earn less, more women look for ways to support their families.

In Chinnamedu, the local NGO "S" runs a primary school and provides free education to the children of fishers. Because of this, it is more popular among fishers. The NGO continues to have a harmonious relationship with uur panchayats and fishers. During the fieldwork, I delved deeper into this topic to examine whether this support to the NGOs had resulted in promoting the political capital of fisherwomen. In Chinnamedu and Chavadikuppam, women's role in fishing is primarily limited to selling fish door-to-door and in both licensed and unlicensed markets, as well as in streets and public places. Many middle-aged and young fishermen of these villages stated unequivocally that they are unwilling to allow their wives to participate in selling and marketing the fish because it would harm their family pride (which is understood as their masculine pride). However, in several instances, they require financial assistance because climate change and the decline in fish catch have severely impacted their income security. Recognizing this, some respondents acknowledged that women's self-help groups provide the option of getting external income into their homes without jeopardizing their family pride. Asset-less (poor fishermen who do not own vessels or equipment) fishermen also noted that they encourage their wives to participate in women's self-help groups as it would support their households during crises. Nevertheless, they are unwilling to permit them to participate in the activities of the local institutions in the fishing villages. Therefore, the fishermen's cooperation with the operation of women's self-help groups is solely to acquire external income resources/credits for their households, but not in political terms. Given these insights, this book demonstrates that the economic participation of women does not have to be reflected in the political institutions of fishing villages. During the fieldwork, I found a handful of recent instances in which fishermen held informal meetings on the beach/public spaces to discuss the eroding coastal spaces, and they collectively decided to draught a petition to build a seawall along their coasts. However, uur panchayats have never considered the opinions and expectations of women regarding seawalls. Notably, women's opinions on seawalls played the least important role in some villages, such as Madavamedu and Kottaimedu, indicating that fishing villages primarily involved in surukku valai fishing have the least concerns about including women's perspectives in coastal development activities and strategies.

The positive impacts of women's agencies on social relations

According to Salagrama and Koriya (2008), women's roles in the fishing sector in Tamil Nadu have diminished much more after the modernization era. This book agrees that this is partially correct. In addition to the insights of Salagrama and Koriya (2008), this book indicates that with the increasing number of women's self-help groups and uncertainty about fishers' income and livelihoods in the face of many stressors (including climate risks and change), the leaders of women's self-help groups have started volunteering to serve as the principal mediators between the fisherwomen and the outside world. In financially unstable fishing villages, women's self-help groups are more active in fishing communities, and women are keen to borrow money from their credit networks. Self-help groups put fisherwomen to work with other women. In this respect, it is also worth noting that women's engagement in fishing communities has not done much to address social inequalities. The Scheduled Caste women of Koozhayar hamlet were excluded silently from the village's mainstream fisherwomen. In the study communities, fisherwomen are confined to their villages and occupations. Fishers are becoming more connected within and across the fishing villages of South India due to the growing boom of information technology. Self-help groups inspire and galvanize women to participate in social activities. Women's self-help groups in Chavadikuppam and Chinnamedu spread information about reproductive health, sanitation, and child health with the help of a local NGO.

This study identifies that the SHGs perform in small fishing villages (such as Koozhayar, Kodiyampalayam, Chinnamedu, and Chavadikuppam) better than the others, though these villages struggle to receive decent credit support from the self-help groups in times of crisis. The fisherwomen of these villages perceive women's self-help groups as a great way to get low-interest loans. There has been an increase in women's participation in self-help groups in small fishing villages. In capital-intensive fishing villages, it didn't happen. The best example is Madavamedu. According to the accounts of Madavamedu's women leaders (including the secretary of women's self-help groups), uur panchayats must be more gender-sensitive and lenient in allowing the activities of women's self-help groups without imposing unnecessary restrictions. Twenty young and middle-aged fishermen were interviewed and subjected to focus groups to gain insights on this topic. They never held a formal or informal position in the villages – they were asked to share their views on the roles and contributions of women's self-help groups. Surprisingly, roughly half of those fishermen knew nothing about women's self-help groups or their activities beyond the possibility of receiving credit support for their households. The remaining respondents had some rudimentary understanding of the benefits of SHGs. It was repeatedly observed that fishermen held similar views about women's self-help groups, namely that neither women nor women's SHGs could effect radical and significant changes in the lives of fishing populations. However, interference by individual fishermen in the activities of women's self-help groups was extremely rare. As a result, it can be deduced that it is not individual fishermen who are opposed to the roles that women play in local institutions, but rather uur panchayats. Taking this speculation forward, this book claims that it is not the individual fishermen but the local institutions that make it hard for outside actors to work in the tsunami-hit fishing villages of South India.

The changing trends – a comparison

This section aims to draw parallels between the study on the effects of traditional governance systems in Mexico and the evolving local governance systems in the fishing villages across the study area (Diaz-Cayeros et al., 2014) as both the cases share certain similarities. Mexico underwent gradual democratic processes in the 1990s after the Presidential elections. Before 1995, Mexico had both traditional indigenous and formal (political party) governance systems. Oaxaca, a state in Mexico, enacted reform in 1995 that gave *usos y costumbres* (traditional governance systems) legal standing. It formally acknowledged the right of indigenous people to choose their form of government and permitted indigenous communities to choose between "traditional indigenous governance systems" (*usos*) and "party-based governance." The authors (Diaz-Cayeros et al., 2014) compared the effectiveness and public and political participation of the two governance systems and found that traditional governance (*usos*) does not impede democratic development. In contrast, traditional institutions increase public and political participation and access to public services.

There are some striking parallels between the traditional governance systems of the marine fishing villages along the Nagai-Karaikal Coast, where the subject of this study is located, and Oaxaca, a state in Mexico. Both were governed by formal and informal (customer-based) structures, which ran in parallel. Traditional forms of government were well received and accepted by the populace in both the regions. Both traditional systems excluded women from local decision-making and participation. Diaz-Cayeros et al. (2014) observe that, since the implementation of the conditional cash bi-monthly transfer (Oportunidades) program in 1997, discrimination patterns against women have shifted in recent years. Then, women began contributing to the education and health sectors, although their contributions were insufficient. This book, on the other hand, shows that after the 2004 Indian Ocean Tsunami, gender bias changed mostly because international donor agencies and NGOs got involved. This made women's self-help groups in the fishing community more effective. Women were asked about their awareness of fisher livelihood challenges. In most cases, except for a few, women responders lacked understanding of the outside world and current fisher issues. However, this is unrelated to climate change and its effects on fishers and their means of livelihood. Government televisions facilitated the flow of information while increasing people's interest and self-effort to assist their households. Results strongly indicate that frequent meetings held by women's self-help groups have indirectly aided in the enhancement of their capacity-building processes, hence indirectly contributing to the fishing village's overall climate adaptation efforts.

Conclusions

In sum, this chapter demonstrates how economically stable fishing villages of tsunami-hit fishing villages on the Coromandel coast of South India maintain rigid social relations and how their local institutions are blind to gender issues, thereby the rise of weakening trends of women's agency. In contrast, economically disadvantaged small fishing villages recognize the inherent income certainty and welcome external agencies with open arms, refraining from imposing restrictions. The result is the proliferation of women's self-help groups. The best examples are Chinnamedu and Chavadikuppam. In these villages, women's self-help groups function more smoothly than others. In recent years, these groups began training women to recognize their rights and the need to contribute to their fishing households' economy.

Field et al. (2010) identify that scheduled caste women face fewer restrictions than upper caste women. Women, even upper caste women, suffer from stringent social norms that inhibit entrepreneurial aspirations, yet they respond to "external" training and support. Although the situation is different, the prima facie observations of this study (Field et al., 2010) confirm the claims made in this book about social norms and restrictions on women. The fishing villages of this research are largely homogeneous (single caste: Pattinavar), with two villages (Koozhayar, Kodiyampalayam) having a smaller but significant scheduled caste population. Discussions with those Scheduled Caste women show that they face fewer restrictions from men and local institutions than Pattinavar fisherwomen. This indicates that, unlike the scheduled caste community, Pattinavars possess rigid social norms and are tightly governed by traditional institutions. Though the benefits of these norms in fisheries management and developing fishers' social capital are widely documented, these social norms indirectly and directly restrict the goals of fisherwomen, hampering their efforts to bring additional revenue into their households. Along with that, in this context, it is equally important to highlight the ways in which household wealth and financial capital of the fishing villages restrict women in their regions in diverse ways. During the fieldwork, it was observed that the women of small fishing villages that possess limited financial capital are subjected to fewer gender-based restrictions. One explanation for these limited restrictions is that the fishermen of these villages anticipate loans from self-help groups. So, they are collectively lenient in imposing restrictions on the enterprise of women. Another explanation is that the uur panchayats learned painful lessons from the state government's flood compensation measures and realized the importance of the support of NGOs and other external agencies for them to negotiate with the state government to receive their benefits, including compensation for climate risk. These two hypotheses are based on anecdotal evidence gleaned from field observations rather than solid empirical research. Furthermore, these explanations depend heavily on context and cannot be generalized across coasts. Recent scholarship (Gangadharan et al., 2018) provides empirical evidence that female leaders are equally and significantly more likely to act as deceptive leaders. Because of this, it is not enough to simply promote the activities of women's self-help groups and the influx of external agencies into the fishing villages to foster overall inclusive growth. The true necessity is to attempt to alter the social environment in which male and female leaders function.

This chapter highlights the necessity of the mutual efforts and collaborations of the state and NGOs to construct a disaster-and climate-resilient fishing community. The microcredits of women's self-help groups to the fishing households have been beneficial to them during rough seasons and fishing bans, and they promote women's agency. Discussions with NGO staff and women self-help group members reveal that external agencies (NGOs and fund donors) need to understand the socioeconomic context of fishing villages and the context of their vulnerability. In practice, women's self-help groups and their fund donors, that is, NGOs, operate on the lines of their operational rules but not in accordance with the vulnerabilities of fishers. Such approaches are inadequate and mismatched to address the unique vulnerabilities of fishing communities brought about by climate change. It reduces the adaptive capacity of the fishers and, in some cases, it induces maladaptation. We need more research, and qualitative insights from the fishing villages across this stretch are required to understand the capacity building and potential of self-help groups to incorporate adaptive climate planning. At the same time, there are huge knowledge gaps in these topics to determine whether microcredit has empowered women and increased the resilience of the fishing community's livelihoods or whether it has been instrumental in reducing the vulnerability of fishing households to climate change and associated risks. The data must be context-specific and co-produced with fishing community stakeholders to identify their needs, expectations, potential, and constraints (Ayeb-Karlsson et al., 2015).

As Ray-Bennett (2010) suggests, the catalyst organization must constantly monitor the SHGs to comprehend and evaluate the effectiveness of the members during the post-disaster and pre-disaster phases. Ray-Bennett warns that microcredit alone cannot eliminate women's susceptibility to natural disasters, and she advocates an integrated, collaborative, and instrumental approach among the state, the market, and civil society. The state must take on multiple roles for women's self-help groups and the fisherwomen cooperative society to flourish in the fishing villages across the Coromandel coast of South India. To begin, local institutions that are solely meant for women require effective rules, regulations, and functions to operate. For example, we require a clear distinction between the tasks that fisherwomen cooperative societies are expected to perform. During the fieldwork, it was repeatedly observed that the shared distribution of tasks and resources between fishermen and fisherwomen cooperative societies rendered the latter weak and defunct and diminished women's agency over the period. Using Tamil Nadu's fisheries documents as evidence, this book indicates that the state's fisherwomen cooperative societies were poorly conceived and executed. In the most recent fisheries policy note (Government of Tamil Nadu, 2018, 2021, 2022a), the roles, responsibilities, visions, and financial resources of fisherwomen's cooperatives receive the least attention. There is a pressing need to clearly define the roles and aspirations of fisherwomen cooperative societies, and this document needs to be regularly updated to reflect the expanding body of relevant literature. Second, dedicated committees at the district and taluk levels are required to track and evaluate the progress of fisherwomen cooperative societies. The committees need to have a better understanding of the vulnerability and socioeconomic context of the fishing villages, but this does not require a top-down administration structure. The committee can increase women's involvement by consistently backing fisherwomen cooperative societies. Third, fisherwomen cooperative societies should have their own offices at the hamlet level, with regular oversight from the district administration. Such regular interactions will help them build networks and form partnerships. As discussed in the last chapter (Chapter 6), fisherwomen cooperative societies and local NGOs must work together to build mutually beneficial relationships and share resources. In order to ensure that these collaborations provide avenues for designing and implementing disaster management and climate action strategies at the village levels, the state fisheries can set up a separate committee to oversee them. Such partnerships can provide collaborative training and assistance to women's self-help groups and fisherwomen cooperative societies to respond to disasters and climate extremes. Fourth, and most importantly, the state fisheries department needs to ensure constant material and non-material support to the fisherwomen cooperative societies through the appropriate channels, which is crucial in building a resilient fishers' community that can withstand and fight against climate change and disasters. Overall, this chapter presents a mixed bag of field findings. Local governance systems of fishing villages are dominated by the heads of uur panchayats, calling into question the adequacy of participatory spaces for women in these systems. Though there is no gender oppression, the restriction against women is subtle and structured. Bavinck, a noted scholar of the maritime fishing communities of South India's Coromandel coast, contends that gender sensitization in the Pattinavar fishing communities should occur naturally rather than through external intervention. Otherwise, it may result in undesirable effects (Bavinck, 2006). Given the Pattinavars' historical and cultural background, we cannot expect radical changes in their fishing society. In recent years, the interactions and local governance systems of the fishers' local institutions have advanced significantly due to globalization and technological advancements. It is reasonable to anticipate more progressive steps in the future.

Notes

1 Esther Fihl, an anthropologist and scholar of Cross-Cultural and Regional Studies, has four decades of ethnographic experience in the fishing villages of the district of Nagapattinam. In a personal communication, she shared the raw data (transcribed file) that was collected by Esther Fihl (with the help of a translator) in 2010 from the women respondent and the uur panchayat leader of a fishing village in Nagapattinam district. The transcribed file gives detailed insights about Tharangambadi uur panchayat cases. Notable elements described in this interview suggest that the uur panchayat is authoritarian and gender insensitive in handling matters involving women's problems (Fihl, 2010, Personal Communication).
2 While the government of India frequently employs the vague term "microfinance," Sundar (2010) casts doubt on its practical application. The terms "microfinance" and "microcredit" have many functional differences. NGOs are microcredit organizations that provide and recover loans. In contrast, "microfinance" is an umbrella term for a vast array of activities, including savings, loans, insurance services, community empowerment, and social mediation. Microfinance activities in the community can be carried out by formal (state, cooperative banks, etc.), semi-formal (NGOs, donor agencies), and informal (money lenders, etc.) institutions (Qudrat-I Elahi & Lutfor Rahman, 2006). The present study demonstrates that the functions of women's self-help groups are primarily limited to loan distribution and collection. Consequently, as Sundar (2010) notes, "microcredit" is correct when referring to these loans. In a few small fishing villages, however, the activities of women's self-help groups expanded to include organizing the community to respond to climate change. Given these considerations, I prefer to use the terms "microcredit" and "microfinance" interchangeably throughout this chapter and according to the context.
3 Here, I purposefully use an alphabet to refer to this NGO to not reveal the identity.
4 It should be noted that this narrative was briefly pointed out in the previous chapter (Chapter 6). In order to reemphasize the issue at the ground levels, I have elaborately documented this narrative in the present chapter.
5 Here, I purposefully use some alphabets to refer to the NGO and women self-help groups to not reveal the details of the particular organizations.

8 Conclusion and the way forward

Fishing technology and exports dominate global fisheries research, leaving out the primary stakeholders of the fisheries sector, who are small-scale fishers and fish workers, including fisherwomen. This is not to say that fishers' history, rights, and governance systems have not been explored. There is a steadily expanding body of anthropological and legal anthropological literature devoted to these topics in the context of Tamil fisherfolk (Subramanian, 2009; Subramanian, 2011; Bavinck, 2011; Bavinck & Jyotishi, 2014). However, there is a serious gap in the expanding literature on climate change worldwide due to the paucity of social science scholarship in coastal India, including on the Coromandel coast of Tamil Nadu, that links marine fishing communities with climate change and adaptation. Detailed case studies on local perceptions, local knowledge, vulnerability, adaptation, and local institutions of marine fishers focusing on the Coromandel coast are particularly difficult to find, despite the fact that these regions are among the most vulnerable global hotspots for climate change and have the potential to offer climate change lessons and prompt scholars, practitioners, and policymakers to reexamine the existing body of knowledge. This book, written primarily from the context of coastal Tamil Nadu, not only addresses these knowledge gaps by taking a bottom-up approach to adaptation to climate change but also offers valuable lessons and conceptual insights on this topic to a global audience. One of this book's important contributions is its proposal of a unified framework that combines the sustainable livelihoods framework and the AIL framework, allowing readers to investigate the local adaptation of fishers to climate change through their capitals and institutions. This approach provides a lens through which to examine the functions, potentials, and complications of marine fishing communities in responding to climate change, and it is easily scalable to other geographical contexts. Backed with empirical insights and local voices, this book emphasizes social and political capital for successful and sustained adaptation to climate change. Although this study focuses on a single Tamil Nadu district, it connects to the vulnerability and adaptation of fishers from other coastal regions of the nation and the world. Comparing the empirical findings of this study with those of others yields a full comprehension. Adapting "locally" is a crucial component of the solution to the global problem of climate change, which has an impact on people

DOI: 10.4324/9781003187691-8

all over the world. This book seeks to bridge the divide between academic, practical, local, and global perspectives on climate change.

The focus of this study, the Nagapattinam district on the Coromandel coast, has several vulnerabilities to climate change and coastal disasters. Small-scale fishers in this region have been hit hard by overfishing, coastline erosion, government disinterest, and market swings in recent years. Hence, the lives of the struggling small-scale fishers along this coast are already being negatively impacted by climate change and multiple complications, which is now firmly shown by a rising body of scientific evidence in recent decades. To make matters worse, the long-term effects of the 2004 Indian Ocean Tsunami continue to afflict them, with far-reaching ramifications for the centuries-old internal governance structures of the Pattinavar villages. Against this background, this book addresses this question: how do marine fishing communities along the Coromandel coast adapt locally to climate change? The chapters of this book turned this single question into four themes and were interrogated in the study setting. (i) local indigenous knowledge and climate perceptions, (ii) vulnerability and adaptation of marine fishers to climate change, (iii) local institutions and localized adaptation, and (iv) women's agency and climate change adaptation (CCA). The outcomes of this study, primarily localized adaptation interventions of the marine fishers and discussions on the local institutions of fishers, can readily be extrapolated to comprehend the other marine fishing communities of Tamil Nadu and the neighboring states of Kerala, Andhra Pradesh, Karnataka, and Odisha. This is because small-scale fishers in India, particularly, in South India, face almost identical risks to their livelihoods both in terms of climate change and the growing developmental complications, and resource politics. In recent decades, the repercussions of climate change have been evident on the coasts of South India. Still, I warn that the setting of this study should be taken into account when extrapolating the results of this work to other research regions and settings. Internal (local) government systems of fishing villages throughout the south Indian coast show remarkable parallels as well as variations, which is why this is so important. The sections that follow give a summary of the study's findings as well as policy recommendations to bolster CCA at the local level.

Indigenous knowledge and climate change perceptions

In India, the local ecological (indigenous knowledge) systems of marine fishermen have been the focus of a growing corpus of fishing research. Some select anthropological works: Hoeppe, 2007 (Ecological knowledge of the fishers of South India); Kodirekkala, 2018 (Cultural adaptation and indigenous people of South India). On the other hand, there is a limited grasp of the social dimensions of indigenous knowledge systems and climatic perceptions of maritime fishing communities in the face of climate change. It should be underlined that none of the existing studies focused on how small-scale marine fishermen, the primary stakeholders in fishing, perceive and depend on their indigenous knowledge systems in the context of climate change. Knowledge of the ways in which fisherwomen perceive climate change in the context of their own indigenous (local ecological) knowledge systems is even scarcer in the sparse literature. Therefore, in the context of India, further research is required to grasp clearly how both fishermen and fisherwomen perceive the escalating threat of climate change. To make sound climate change policy, the perspectives of fishers (particularly fisherwomen) are critical. In-depth regional studies are crucial for understanding the inconsistencies and ambiguities in stakeholders' beliefs and actions about climate change, as supported by previous literature (Lorenzoni et al., 2007). Exploring indigenous knowledge systems and climate perceptions of fishermen can pave the way for a rapid examination of their coping and adaptation responses to climate change.

Most fishing respondents in this study are concerned about the shifting weather and climate patterns, especially after the 2004 Indian Ocean Tsunami disaster. According to their accounts, rainfall patterns have become variable and unpredictable over the past few decades, and they linked these changes to their fishing livelihoods in numerous ways. Across the world, climate change is predicted to bring more hazards and natural disasters in the future, and growing scientific evidence links the increasing intensities and frequencies of climate change to extreme compound events and associated disasters (Anderson & Bausch, 2006; Bouwer, 2011; AghaKouchak et al., 2020). Whereas to understand the interconnectedness of these events, we need more in-depth research; in the Bay of Bengal, the knowledge gaps concerning local-regional climate events and the interconnections of temperature rise to atmospheric and oceanic extremes are glaring. More research is needed to determine the specific ways in which global warming raises the risk of tsunamis in the future. Overall, the climate change perceptions of respondents have five key manifestations: (i) Erratic and unpredictable climate patterns and the resulting seasonal shifts over the decade, (ii) increasing frequency and intensity of extreme weather conditions, (iii) coastal erosion, (iv) the continual decline in fish capture and species decline, and (v) the long-term consequences of the 2004 Indian Ocean Tsunami disaster: At first glance, the fifth manifestation appears out of place in discussions about climate change, but it is not. Though climate change and tsunami disasters are the outcomes of different factors, both hit the fishers across this coast. In most cases, climate change effects mix with the long-term consequences of tsunami disasters, making small-scale fishers more vulnerable and puzzled. There is ample evidence that the 2004 tsunami had huge impacts on marine fishers of Tamil Nadu, and it continues to harm them in multiple dimensions (Murugan & Durgekar, 2008; Arlikatti & Andrew, 2012; Shanthi et al., 2017; Swamy, 2018). Across Asia, the complex interactions between climate change and the long-and short-term effects of coastal disasters require more exploration, and we also need to look more closely at how climate change effects increase the frequency and intensity of coastal disasters, including tsunami disasters. This research should be directed toward the coasts of South India. This book provides a solid introduction to these themes by highlighting the voices of tsunami-hit maritime fishers of South India.

The empirical findings of the present study indicate that indigenous knowledge is becoming less popular among younger fishermen compared to older fishermen for two reasons. Young fishermen rely heavily on weather and climate forecasts provided by various media outlets (TV, mobile phones) to keep up with the latest developments. However, this information is primarily limited to major climate-related risks and excludes minor ones. Despite their reliance on "science," fishermen frequently encounter difficulties spotting potential fish shoals and erratic sea currents. For this, they also look to the government and non-profit organizations for assistance in providing regular and reliable weather updates. Most senior fishermen in the area, on the other hand, rely on their traditional knowledge systems for their livelihoods. Because so many fishermen are leaving, the indigenous knowledge of elderly Pattinavars is not being passed on to the next generation properly, and this trend will only likely worsen in the future. In addition, fishermen are becoming less reliant on indigenous knowledge systems because they are moving abroad for work, using TV and cell phones more, and getting help from outside sources like MSSRF and INCOIS (Samuel, 2016).

Small-scale fishing households on the Coromandel Coast rely heavily on marine fishing. Already, the mechanization of fishing has made their way of life more fragile and difficult. They have no option but to rely on fishing as their major means of survival and income. This book demonstrates that environmental and climatic changes are becoming increasingly harmful to the well-being of maritime small-scale fishing households in these regions. To confront these multiple challenges posed by climate change, it is vital to building a state-level organization that can provide reliable meteorological information to fishermen, particularly at the taluk level (which can be expanded to the village level in the future). By collaborating with a non-profit organization, the Indian National Centre for Ocean Information Services (INCOIS) has already initiated efforts to give scientific help to coastal fishermen in Tamil Nadu (including the district of Nagapattinam) through multiple channels. It gives mobile app features to fishermen, such as fishing zone weather updates. Despite this, the results show that fishers in the research area have limited access to scientific support from outside sources, which is inconsistent in some cases. To help reduce the burden of small fishermen, scientific updates about potential fishing zones and fish shoals can be helpful. Fish caught by small-scale fishermen will grow in size, and the risks of fishing will be reduced as a result (NCAER, 2010). A wireless communication network and livelihood assistance from the World Bank have also been granted to help Tamil fishermen feel safer and more secure in their employment (Government of Tamil Nadu, 2018). Efforts are being made to reduce the susceptibility of small-scale fishers to climate change impacts and to strengthen their adaptive capability. The state administration and other authorities may consider expanding the program by targeting small fishing villages engaged in nearshore fishing. To improve their adaptive capacity, the state could consider providing small-scale fishers with climate change education and disaster management training. In light of the findings, this study does not diminish small-scale fishers' indigenous knowledge systems and their historical ability to deal with and adapt to numerous stressors, such as climate risks and changes. Instead, it focuses on the significance of understanding the limitations of indigenous knowledge systems of marine fishermen in foreseeing and addressing the effects of climate change. Even two decades ago, Kandlikar and Sagar (1999), focusing on the case of India, emphasized the importance of examining more the political and cultural perspectives of climate change, as well as the need to develop more knowledge and own "models" for participating in and shaping effective international climate change policy discussions. Agreeing with this, this book advocates more research into how fishermen adapt to climate change by leveraging local knowledge systems and traditional traditions because local is congruent with adaptive management. Furthermore, this book underlines the need to understand the shifting transitions and aims of small-scale fishing communities in the current era by focusing on a few selected accounts from the field.

Vulnerability and adaptation to climate change

Four essential points on the climate change vulnerability of Pattinavars' fishers have emerged from this study, which are as follows:

i. Fishers (particularly fishermen) perceive climate change consequences in increasing temperatures, decreasing rainfall, irregular and turbulent weather patterns, falling fish populations, shrinking fish stock, coastal erosion, rising groundwater salinity, and increasing cyclones over the last three to four decades.

ii. Climate change exacerbates Pattinvars' vulnerable livelihoods, which are already threatened by overfishing. It worsens their income insecurity. In some instances, changes (including climate change) have also given possibilities to fishers (Macchi et al., 2015). In this context, it is important to highlight the works of Michael H. Glantz (Glantz, 1988; Glantz, 1995), who introduced the concept of climate change "winners and losers" roughly three decades ago. In these works, Glantz argues that climate change clearly divides nations/regions/communities into winners and losers. As he correctly stated, in the future, some (nations/regions/communities) will benefit from climate change while others will suffer. This study corroborates Glantz's insights (1988, 1995). In the fishing villages, it was also observed that climate change provides some households with benefits, as it prompted them to consider occupations other than fishing. Except in a few cases, non-participant observations throughout the study sites demonstrate that fishermen's migration has greatly improved fishing households' financial security. Such income security supports fishing households during and after climate risks.

iii. In many instances, fishermen were bewildered about how to respond to erratic weather patterns and the concomitant weather and climate impacts. They have begun to recognize that their prior experiences with coastal ecosystems and climatic patterns are no longer relevant in the face of changing environmental conditions.
iv. Many factors shape, influence, and worsen small-scale fishermen's sensitivity to climate hazards and change – (a) the geographical isolation of the fishing villages, (b) poverty, (c) poor infrastructure, such as bad housing, lack of roads to nearby towns, and cyclone shelters are not up to par, (d) less public transportation, (e) poor sanitation and wastewater management, (f) seawater intrusion and the increasing salinity levels, (g) the apathy of government and bureaucracy for small-scale fishers in administering compensations, (h) rising inflation, and (i) weak political and human capital.

Kelman (2014) warns that over-focusing on climate change may depoliticize local communities' developmental challenges, shifting the focus away from reducing their real vulnerabilities and further marginalizing them. In agreement with the caveats and arguments of Kelman (2014), this book highlights the developmental and infrastructure challenges that Pattinavars face and demonstrates that there is an urgent need to address them in order to reduce their vulnerability.

As we can see, there is a clear analogy between Pattinavars and marine fishermen around the Indian coastlines regarding the four key points of Pattinavar vulnerability and adaptive responses. Small-scale fishermen in India often put their lives in danger to support their families, which makes them more vulnerable to climate change and severe weather. Purse and ring seine fishers in better economic positions risk their lives to protect their fishing assets from climate catastrophes since they are doubtful of receiving compensation from the state government. So, fishers' livelihoods are continually in question, notwithstanding the state's social support systems and laws. Insufficient income, unstable finances, and a lack of proper disaster education exacerbate the vulnerabilities of fishers to extreme climate conditions. Fishers with stable finances can better respond to climate events (OECD, 2010; Stoll et al., 2017). Stoll et al. (2017) write that more research is needed to determine how financial capital, community connections, and individual expertise influence fishers' adaptability to climate change. Such factors could harm the fishing community's adaptive capacity. "Migration" has emerged as a potential coping strategy for fishing households to respond to income uncertainty, which has increased their resilience to climate change. Many scholars worldwide have written about fishers' migration (Badjeck et al., 2009; Islam et al., 2014b; Vivekananda et al., 2014; Shaffril et al., 2017). Besides climate change, many social, political, and economic factors influence the migration of fishers. As this book shows, the migration of fishermen has given their fishing families a lot of new opportunities and changed the way marine fishers all over the region live.

Growing migration among young and middle-aged fishermen over the past few decades has led to a labor shortage in the fishery (Jacobson et al., 2019), and in the case of the Coromandel coast, it is one of the factors among the young fishermen for the loss of emotional attachment to their ancestral occupation. As highlighted earlier, it has also decreased their reliance on indigenous knowledge systems, which may affect the district's small-scale fishery in the long run. Despite the adverse effects of migration, field observations show that migrant remittances have ensured the food security and well-being of fishing households in the face of climate change. Like this, there is little knowledge of CCA barriers that exist in India's fisheries. No empirical study has analyzed the obstacles and limitations faced by small-scale fishing communities in Tamil Nadu in adapting to climate change. Recent empirical work in Bangladesh by Islam et al. (2014b) finds that formal institutional barriers like difficulty obtaining credit, unfavorable credit schemes, and difficulty accessing fish markets limit the fishing community's ability to adapt to climate change. In the context of the present study, this book puts forward the "apathy of state institutions" as one of the barriers that obstruct the well-being of the marine fishers of not just Nagapattinam but also the entire coasts of South India, which could negatively affect the resilience of marine fishers in adapting to climate change and associated risks.

In a study on institutional barriers in Sri Lanka, Panditharatne (2016) finds that a lack of monitoring and evaluation is a major drawback in adaptive decision-making. Similarly, this study finds that the state government of Tamil Nadu has already established nodal bodies, provided numerous welfare programs for fishermen, and allocated substantial funds for a comprehensive flood and cyclone protection strategy to respond to disasters (TNSDMA, 2016, 2018). On the other hand, the fieldwork of this study reveals that, in the recent past, the bureaucrats rarely invited the perspectives and engagement of the local fishing communities in the discussions and implementation of projects on sea-wall construction. This is extremely limited in the case of small fishing villages, showing there are only limited and scattered dialogues between the upper levels of the governance systems and local institutions. As shared by the heads of the local institutions, small-scale fishers are largely ignored in the debates on sea-wall construction on their coasts, and the majority process has been a top-down approach. Even if the state invites the local leaders, the invitation is mostly selective and not on the lines of the existing range of local institutions in the fishing villages, showing that there has been an inadequate institutional linkage between the upper and lower echelons of the institutions in Tamil Nadu's maritime fishery. To manage this, to begin with, the local institutions should identify the active "local champions" who are exclusively meant for discussions related to climate action and disaster management. The local institutions need to assign them "roles" and "tasks" to be met, and they are supposed to serve as the authentic negotiators between the state and the fishing community. In the future, those local champions can adequately represent the fishing community's voices to the state government. Given these reasons, this book argues for engaging uur panchayat leaders in adaptive decision-making and co-management. This book demonstrates the ways in which fishers' social capital, including their efficient social networks and political involvement, helps reduce their vulnerability and bounce back from climate risks. Scholars worldwide have held similar discussions (for example, Adger, 2003; Badjeck, 2008; Jordan, 2014; Cinner et al., 2018). In addition, the findings of this study reveal an interesting fact: the poverty of small-scale fishers "motivates" them to break out of rigid cultural practices and become "flexible" in approaching external agencies (including NGOs and the fund donors who support women's self-help groups), which provide income opportunities for fisherwomen. According to Cinner et al. (2018), "flexibility" – a crucial adaptive capacity domain – influences the capability of individuals and systems for climate adaptation. In light of this, the empirical insights of this work indicate that fishers that are adaptable enough to accommodate external actors are better able to cope with and adapt to climate-related impacts. The "flexibility" of Pattinavar fishing institutions allows external actors to work for the "good" cause of fishing villages, increasing their adaptive capacity to climate change. For instance, the local institutions oppose letting Pattinavar fishers switch their fishing occupation and cause conflicts with the external actors for multiple reasons, reducing the overall adaptive capacity of the fishing villages, which is highlighted in this book.

Local institutions and localized adaptation to climate change

Uur panchayats are the most notable local entities on the Coromandel coast. Because of this, earlier research concentrated primarily on the capabilities of ur panchayats while ignoring the roles that other local institutions play in disaster management. Local institutions like fishermen cooperative societies and self-help groups in coastal Tamil Nadu can play crucial roles in coordinating fishers and maintaining fisheries' livelihoods amid climate stress since they are already well connected with fishing communities and external players in the region. Furthermore, these institutions play critical roles in safeguarding the social safety nets of fishers throughout the year (Kurien & Paul, 2001) and food security. Customary (uur panchayat) and formal/semi-formal (local administration, fishermen cooperative association) institutions organize and administer fishing communities differently. As a result, this book also documents the ongoing power struggles among local institutions, which impacted the adaptive responses of the fishing communities. The circumstances in which these power conflicts occurred are complex and multifaceted. These power conflicts among local institutions impede the adaptive capacity of small-scale fishers in Tamil Nadu, which has yet to be taken into account by the state government's policy notes on disaster management and fisheries policy. The study's fieldwork was targeted and selective, and it did not intend to provide robust empirical evidence to support the argument – "power contests" among local institutions and their immediate effects on adaptive decision-making along the coast of Tamil Nadu. However, as some specific cases from women respondents' accounts suggest, uur panchayats' overreach into women-centric local institutions (women's self-help groups and fisherwomen cooperative societies) has resulted in the formation of the namesake, feeble, and quasi-representative local institutions. All of these have had visible impacts on the overall adaptive capacity of the fishing villages in responding to climate change and risks. There is a compelling need for more regional research on the coasts of South India to identify the winners and losers of local climate adaptation actions. To continue and advance disaster risk reduction and climate change awareness, external agencies should work in conjunction with a variety of local institutions in Tamil Nadu's fishing villages. There is a pressing need for multi-level coordination and subnational action in climate change planning in Tamil Nadu with the state government's effective assistance. As this book advocates, in addition to the uur panchayats, the political parties are the most influential local institutions in representing fishers to external agencies. In recent decades, much has been written about South Indian marine fishers' legal systems, property rights, and increasing politicization (Halfdanardottir, 1993; Kurien, 2007; Korakandy, 2008; Jentoft et al., 2009). Despite this, no comprehensive scholarships are focusing on coastal India on the ways in which political parties can engage with the fishing communities to respond to climate change and disasters. Previous studies on coastal Tamil Nadu either overlooked or undervalued the role of fishermen's cooperatives in disaster recovery and rehabilitation. By highlighting these knowledge gaps, this book underlines that fishers depend on fishermen cooperative societies for state-sponsored social welfare schemes and many other local assistance needs, which play a critical role during and after climatic risks and coastal disasters in ensuring their well-being. Unlike uur panchayats, fishermen cooperative societies are overseen by the fisheries commissioner and the state government, and at the hamlet level, cooperative societies implement state welfare schemes for local fishers. Hence, they are locally rooted and possess enormous potential to address the needs of fishers during emergency-related situations. Therefore, the state and other external actors should not undervalue its potential when implementing disaster resilience and climate adaptation programs in marine fishing villages. Besides, uur panchayats have limits in making and implementing plans and solving the livelihood problems of marine fishers. CCA is a sociopolitical process that involves local politics and community involvement. Therefore, the public sector needs to have clear goals for local fishers and the heads of all the local institutions along the lines of their functions and capabilities. Instead of reinforcing top-down adaptation, instead of reinforcing top-down adaptation options, policymakers should consider local actors, socioeconomic factors, local politics, power, social inequalities, and political possibilities (Eriksen et al., 2015). Amundsen (2012) suggests that community resources, community networks, local institutions, and connections between people and places all add to a community's resilience and may make it easier for them to adapt to climate change.

All the fishing villages studied have a population of around or less than 2000 people. Since they are less numerous than the surrounding big fishing villages, they continue to have less clout when bargaining with the state and major political parties for their demands, which is clear from the respondents' voices. In contrast, because of their sizeable population, those populated villages have a steady collective financial base, putting them in a strong bargaining position with political parties and leaders. Since most respondents in this study are cognizant of "reality," they have had difficulty seeing eye-to-eye with state government officials and political leaders when bargaining for benefits. Underscoring this, this book argues that fishing communities with strong local leadership can more effectively get their suggestions on the political agenda, which can be reflected in climate adaptation actions in multiple dimensions. In the case of climate events, in response, government officials visit the villages, survey them, and reassure them of compensation. Such actions do not occur in villages that lack political representatives who are aware of the ongoing crisis and are capable of negotiating with the district administration. Whereas the fishing villages, those who do not have influential local leaders are more likely to be left out by the state's bureaucratic machinery in CCA discussions, including the construction of seawalls – which is already happening along the fishing villages of this coast.

Through their social networks and caste ties, Pattinavar fishers, regardless of the fishing village, have a substantial stock of social networks. The adaptive capacity of the fishers is dramatically impacted by the fact that this book indicates a skewed distribution of collective financial and political capital among fishing villages. I should clarify here that I did not do a quantitative analysis to determine the social and political capital stock of the fishers. This book presents a qualitative analysis based on the responses of fishermen and explores the practical experiences of inequitable relationships among fishers who share a similar (almost) vulnerable physical space near the coasts. In the light of fieldwork conducted in the fishing villages of the Coromandel coast of Tamil Nadu, it also explains that the effective social networking, decision-making, and implementation of the local institutions can, directly and indirectly, bolster the adaptive capacity of the fishers to climate change and risk. It demonstrates, via empirical findings, that climate adaptation planning necessitates long-term sustaining actions to build community disaster and climate resilience. For that, the "capitals" of the fishers and fishing villages must be interlinked. This book highlights that despite the strong marital and kinship links, there is a disparity in the stock of "capitals" amongst Pattinavar fishers across the fishing villages. This is true not just on the Coromandel coast of Tamil Nadu but also on other coasts of India. As a result, this book cautions readers against generalizing and overestimating the social networks of maritime fishing communities. Grafton (2005) observes that the social capital of fishing communities may have substantial effects on their trustworthiness, civic engagement, and social networks. Aldrich (2011) writes that bonding and linking the social capital of fishing villages (understood as fishing communities) helped them receive support from domestic and international agencies without complications. The empirical discussions in this book support the arguments put forward by Grafton (2005) and Aldrich (2011). In addition, this book highlights that bonding and linking the social capital of Pattinavar fishing villages along the Coromandel coast is insufficient for designing, implementing, and sustaining CCA actions. By pointing out some specific cases in which even caste and kinship relations among the fishers can hinder the strict implementation of rules against unsustainable exploitation of marine resources, this book shows that it, directly and indirectly, influences the adaptive capacity of the fishers to climate change in the long run. As the narratives of respondents to this study suggest, mere bonding and bridging of the social capital of fishers has been less effective at ensuring their localized adaptation efforts to confront climate change and risks. This is evident in the case of small fishing villages where the potential stock of social networks and mutual trust among the fishers exists. The social capital and social networks of marine fishers can mostly be "productive" to them in adapting to climate change if and only if it is linked to their human and financial capital. The social networks, political capital, and human capital of marine fishing communities influence domestic and international agencies in securing material means of survival. Long-term adaptation to climate change in fishing villages along the coasts of Tamil Nadu and on other coasts of South India can only be sustained if the local institutions are flexible enough to allow state and private agencies to work together and promote marine fishers' livelihood adjustment actions.

Women's agency and adaptation to climate change

As rightly said by Nalini Nayak, a small-scale fish workers' activist, especially based in South India,

> The fact that men and women participated equally in the fishery brought a certain, shall we say, confidence to women in the fishing community. They were part of the economy, they were the ones who brought money back to the family, so they had this status in the family but didn't have status in the community....
>
> (Nayak, 2017)

By reflecting on her views, this book offers empirical insights to indicate that customary local institutions of centuries-old maritime fishing communities are vastly superior at settling domestic conflicts, distributing community justice, fostering civic engagement in decision-making activities, and assuring efficient public service delivery to the community. Nonetheless, the power structures and social norms of the fishing villages, directly and indirectly, discriminate against women and exclude them from public places and the decision-making processes of the internal governance systems, which has severe ramifications for the overall adaptive actions of the fishers in responding to climate change and associated risks. Because of fast socioeconomic shifts in coastal Tamil Nadu over the last few decades, the conditions of "no status" and "low status" women in Pattinavar fishing villages have changed dramatically in recent years. Significantly, fisherwomen have learned to channel their demands through relatively new local institutions such as women's self-help groups. As the fieldwork in the tsunami-hit fishing villages of the Coromandel coast suggests, the fisherwomen are predominately willing to participate in activities that benefit the economies of their households. However, these trends are not necessarily to be reflected in extending their engagement in strengthening the collective actions of the fishing communities. To describe this, in the fieldwork, a few fisherwomen noted it would compromise the integrity of the fishing community. This is seen in the tremendous rains in 2015. Almost all of the fishing homes in the area were impacted by the severe rains in 2015. After this downpour, fisherwomen attempted to use their social and credit networks to get money to support their families. As the study show, fisherwomen are increasingly more interested in joining women's self-help groups and fisherwomen cooperative societies. It implies that "climate catastrophes" and "disasters" encourage women to take part in activities at locally based institutions that are tailored to their needs and have the pliability to accommodate them. However, their commitment to providing income support to their households has not been reflected in their efforts toward domestic affairs and other acts (that benefit both directly and indirectly) related to climate adaptation. Previous studies (Fordham, 1999; Ray-Bennett, 2009) demonstrate that fisherwomen are more exposed to climate hazards and disasters due to the intersection of gender, caste, and class. This is demonstrated by the fact that the heads of uur panchayats are uninterested in letting their members to participate in domestic issues and other fisheries management-related activities. This is especially prevalent in fishing villages where surukku valai fishing is the primary occupation. The fisherwomen of these communities are more exposed to climate change and associated hazards than the surrounding small fishing villages since most of them are unaware of the need of disaster risk reduction efforts and local participation in numerous "possible" ways.

Academic attention to the gender dimensions of climate change and disasters has increased worldwide in recent years (Gupta, 2015; Enarson & Pease, 2016; Racioppi & Rajagopalan, 2016; Buckingham & Masson, 2017). But in India, not much is understood about how women can respond to climate change and what their limits are. South India's Coromandel coast is no exception. Across the Coromandel coast of Tamil Nadu, as observed by Sundar (2010, p. 68) in the Kanyakumari coast, the beach is a male space – notably, in the case of big fishing villages with significant economic capital, the beach is exclusively a male space, and fisherwomen rarely visit the beach (except the fisherwomen who sells the fish) because their everyday activities are restricted to domestic chores and other related household activities. Therefore, as a practical norm, male-dominated local institutions engage in conversations that influence disaster risk reduction and CCA on behalf of their fishing villages and implement such decisions into the institutions' operational plans without consulting with women. It makes fisherwomen oblivious to the significance of understanding disaster risk reduction measures and current and emerging threats. Several studies examined gender inequalities in the rescue, recovery, and rehabilitation efforts following the 2004 Indian Ocean Tsunami. Many works, such as this one, focus on the ways in which "gender-power" linkages – such as gender divisions of labor, insufficient access to land, unequal ownership, cultural difficulties, and caste and gender intersections – enhanced women's vulnerability to the tsunami disaster (Ariyabandu & Wickramasinghe, 2004; Oxfam, 2005; MacDonald, 2005; Pincha, 2008; Juran, 2012). Since the 2004 Tsunami, the tsunami-ravaged coastal districts of South India have been pummeled by four major cyclones and countless intense downpours throughout the course of the past few years. Nevertheless, as this book reveals, regardless of their socioeconomic status, women have not been adequately prepared to face the hazards posed by climate change and coastal disasters. Several of the women who participated in this study are regulars at women's self-help groups, but they admit they are unable to swim and are generally uninformed about the risks posed by climate change and how to manage them. Enarson and Fordham (2001) stress the importance of women working with grassroots groups to find the "vulnerable" and "vulnerable among the vulnerable" in their community before and after risk events. By promoting women's expertise and knowledge, they say:

> The reduction of physical and social vulnerability must become part of the daily activities of women's neighborhood groups, the grassroots organizations they have formed for sustainable and just development alternatives, and professional and educational networks at all levels.
> (Enarson & Fordham, 2001, p. 135)

While highlighting the potentials of prominent local institutions, this book also draws attention to their neglect, which obstructs women's engagement in domestic affairs, resulting in evident setbacks to climate adaptation actions at the local level. The best method to sustain climate adaptation initiatives along the coasts of South India is to engage fisherwomen in decision-making to respond to climate change, which requires expanding fisherwomen's roles beyond post-harvesting, marketing, and social reproduction activities in light of expected climate change impacts. Engaging women in internal governance systems along the Coromandel Coast will take time, and it will require long-term persistence and dialogues between the state and the community. In the short term, the state and external actors can keep pushing local women leaders to take action to make their communities and gender less vulnerable to climate change.

The way forward

Recent years have seen an increase in the amount of literature devoted to the topic of climate hazard crises in Asia (Krüger et al., 2015; Sternberg, 2017; Ahmed et al., 2020). Except for a few works, many of which are macro-level studies concentrating on disasters and the policy aspects of climate change. Consequently, there exist significant knowledge gaps regarding the regional and local impacts, vulnerabilities, and adaptation actions of resource-dependent communities. In the case of coastal India, these gaps are especially large. Multiple aspects of climate change have yet to be addressed at the regional and local levels in the Indian subcontinent. In this last part of this book, I touch briefly on a few themes that deserve more attention. In recent years, scholarships on disasters and climate change have moved beyond mere analysis on analysis on its impacts on households and communities to investigations of foreign policies, conflicts, and cooperation among the nations. Ilan Kelman has written extensively on disaster diplomacy and its ramifications for nations and among nations (Kelman, 2003, 2006b, 2011b, 2016). What role do climate change and coastal disasters play in escalating tensions between Tamil Nadu's numerous coastal regions and other coastal states in India? As a result of climate change and coastal disasters, could we anticipate that disputes between India's coastal communities/villages will increase and grow more complex? Or will future climate change consequences foster collaboration amongst India's already conflicted coastal communities/villages and diverse social groups? Such many big questions concerning conflicts and cooperation due to climate change and disasters have been left unanswered on the coasts of Tamil Nadu and the other coastal states in India. It is hoped that the scope provided by the discussions, arguments, and insights in this book will enable the exploration of these topics in the future. The similarities and differences between disaster risk reduction and CCA are well-documented (Thomalla et al., 2006; Schipper & Pelling, 2006; Tearfund, 2008; Mercer, 2010; Kelman et al., 2015). Gaillard (2010) proposes a strategy for addressing vulnerability, capability, and resilience in relation to climate, disasters, and development. He stresses the existing gaps between the three sectors: climate adaptation, disasters, and development, and argues for the necessity of strong political will to eliminate the gaps and mainstream an integrated approach to strengthen the community's resilience and capability. It's crucial to take a close look at the fishing communities of India's coast and acknowledge the risks they face, but it's also essential to acknowledge the ways in which they can adapt and grow stronger. Exploring the "winners" of climate change in the future is also crucial and fascinating. It is believed that by asking these questions, coastal India and other parts of Asia and the world may learn from each other and find better ways to adapt to the effects of climate change.

Adaptive climate planning cuts across Indian government agencies, the union government, and several international organizations that have already started action at the district level to mainstream climate adaptation and disaster risk reduction. CCA and disaster risk management (DRM) are becoming more linked, and the state is working to combine them (Government of India, 2017). However, in line with Gaillard's (2010) insights, the conclusions of this study, which are supported by extensive empirical data, imply that the state must go far with mainstream adaptive climate planning within the broader context of CCA, DRR, and development and that this requires strong political will at all levels of government. Expanding on Gaillard's (2010) points,

Kelman et al. (2015) argue that adaptation to climate change can only be made if policies and strategies take into account issues of gender, inequality, development, and sustainability. This book backs Kelman et al.'s (2015) claim that fairness in power-sharing and decision-making systems is crucial to sustaining adaptation to climate change. In coastal Tamil Nadu, the disaster management institutions, policies, and strategies are well established. Notably, the Department of Environment of the Government of Tamil Nadu established a separate Tamil Nadu State Climate Change Cell (TNSCCC) in 2014 to engage in various climate change-related initiatives and disseminate climate change information to diverse stakeholders and build capacity for effective climate change governance. Furthermore, in 2014 and 2019, the Tamil Nadu government drafted climate change action plans. However, both versions were criticized for being unclear about their intended outcomes. The newly elected state administration is making more efforts to address seven vulnerable sectors, including coastal area management, and it has emphasized numerous goals that should be achieved by 2030. (The New Indian Express, 2021a). The most recent Draft Tamil Nadu State Action Plan on Climate Change – 2.0 is available on the Department of Environment and Climate Change, Government of Tamil Nadu's website (Draft Tamil Nadu State Action Plan on Climate Change – 2.0, n.d.). A closer examination of the report indicates, however, that the social and political dimensions of climate change in coastal regions, as well as the multiple intricacies of fisheries and coastal systems (participative management, fisheries co-management, effective coastal governance, and institutional thinking in coastal governance systems as discussed by the scholars: Pascual-Fernández, 1999; Chuenpagdee & Jentoft, 2007; Jentoft & Chuenpagdee, 2009; Chuenpagdee & Song, 2012), receive little or no consideration. As a result, institutional thinking is limited in the policy report, and no institution is principally accountable for fostering and advancing district- and village-level climate adaptation actions. Nevertheless, this is not confined only to the state of Tamil Nadu. Madhanagopal and Jacob (2022) discuss the complacency of Odisha and Kerala's climate action plans, noting that there has been less attention on the engagement of marginal and vulnerable populations in CCA actions. As they critique, in climate action plans, neither the social vulnerabilities nor the potentials of socially excluded communities have been given exclusive attention.

Also, as I come to the end of this book, I want to point out how a few and carefully chosen fisherwomen took part in the study. This is due to the fact that they were frequently unavailable due to their domestic responsibilities, and in some instances, they exhibited a lack of enthusiasm in participating in the study due to the sociocultural norms of the fishing communities and the resulting complacency. In most cases, therefore, only interested female respondents, women leaders (informal), and senior fisherwomen from fishing villages participated in this study. Women who participated in the study were predominantly members of women's self-help groups and had a certain level of education. Future research on the Coromandel coast of southern India may take into account the concluding remarks and investigate the "voices" of vulnerable and underrepresented individuals in marine fishing settlements. It is intended that the insights and recommendations of this book will provide governments and international organizations with more crucial points and evidence to promote sustainable fishing livelihoods and increase the institutional resilience of marine fishers to adapt to climate change. The discussions and assertions made in this book regarding CCA will pave the way for additional studies to determine how political parties are involved and how they influence domestic and international organizations in obtaining material and non-material support to make maritime fishing villages more robust to disasters and climate change.

References

Abeygunawardena, P., Vyas, Y., Knill, P., Foy, T., Harrold, M., Steele, P., et al. (2003). *Poverty and climate change. Part 1. Reducing the vulnerability of the poor through adaptation*. United States Washington, DC: World Bank. https://www.oecd.org/env/cc/2502872.pdf

Acker, J. (1992). From sex roles to gendered institutions. *Contemporary Sociology, 21*, 565–569. https://www.jstor.org/stable/2075528

Adam, H. N. (2015). Mainstreaming adaptation in India – The Mahatma Gandhi National Rural Employment Guarantee Act and climate change. *Climate And Development, 7*(2), 142–152. https://doi.org/10.1080/17565529.2014.934772

Adamson, G. C. D. (2015). Private diaries as information sources in climate research. *Wiley Interdisciplinary Reviews: Climate Change, 6*(6), 599–611. https://doi.org/10.1002/wcc.365

ADB. (2012). *India: Tsunami Emergency Assistance (Sector) Project*. Project Number: 39114. Asian Development Bank. https://www.adb.org/sites/default/files/project-document/74573/39114-013-ind-pcr.pdf

Adger, W. N. (1999). Social vulnerability to climate change and extremes in coastal Vietnam. *World Development, 27*(2), 249–269. https://doi.org/10.1016/S0305-750X(98)00136-3

Adger, W. N. (2000). Social and ecological resilience: Are they related? *Progress in Human Geography, 24*(3), 347–364. https://doi.org/10.1191%2F030913200701540465

Adger, W. N. (2003). Social capital, collective action, and adaptation to climate change. *Economic Geography, 79*(4), 387–404. https://doi.org/10.1111/j.1944-8287.2003.tb00220.x

Adger, W. N. (2006). Vulnerability. *Global Environmental Change, 16*(3), 268–281. https://doi.org/10.1016/j.gloenvcha.2006.02.006

Adger, W. N., Agrawala, S., Mirza, M. M. Q., Conde, C., O'Brien, K., Pulhin, J., Pulwarty, R., Smit, B., & Takahashi, K. (2007). Assessment of adaptation practices, options, constraints and capacity. In M. L. Parry, O. F. Canziani, J. P. Palutikof, P. J. van der Linden, & C. E. Hanson (Eds.), *Climate change (2007): Impacts, adaptation and vulnerability. Contribution of working group II to the fourth assessment report of the intergovernmental panel on climate change* (pp. 717–743). Cambridge, UK: Cambridge University Press. https://www.ipcc.ch/site/assets/uploads/2018/02/ar4-wg2-chapter17-1.pdf

Adger, W. N., Arnell, N. W., & Tompkins, E. L. (2005a). Adapting to climate change: Perspectives across scales. *Global Environmental Change, 15*(2), 75–76. http://dx.doi.org/10.1016/j.gloenvcha.2005.03.001

References

Adger, W. N., Dessai, S., Goulden, M., Hulme, M., Lorenzoni, I., & Nelson, D., Naess, L. O., Wolf, J., & Wreford, A. (2009). Are there social limits to adaptation to climate change? *Climatic Change*, *93*(3), 335–354. https://doi.org/10.1007/s10584-008-9520-z

Adger, W. N., Hughes, T. P., Folke, C., Carpenter, S. R., & Rockstrom, J. (2005b). Social-ecological resilience to coastal disasters. *Science*, *309*(5737), 1036–1039. https://doi.org/10.1126/science.1112122

Agarwal, B. (2000). Conceptualising environmental collective action: Why gender matters. *Cambridge Journal of Economics*, *24*(3), 283–310. https://doi.org/10.1093/cje/24.3.283

Agarwal, B. (2001). Participatory exclusions, community forestry, and gender: An analysis for South Asia and a conceptual framework. *World Development*, *29*(10), 1623–1648. https://doi.org/10.1016/S0305-750X(01)00066-3

Agarwal, B. (2002). *Are we not peasants too? Land rights and women's claims in India*. SEEDS no. 21. New York: Population Council. https://knowledgecommons.popcouncil.org/cgi/viewcontent.cgi?article=1033&context=departments_sbsr-pgy

AghaKouchak, A., Chiang, F., Huning, L. S., Love, C. A., Mallakpour, I., Mazdiyasni, O., Moftakhari, H., Papalexiou, S. M., Ragno, E., & Sadegh, M. (2020). Climate extremes and compound hazards in a warming world. *Annual Review of Earth and Planetary Sciences*, *48*, 519–548. https://doi.org/10.1146/annurev-earth-071719-055228

Agrawal, A. (1995). Dismantling the divide between indigenous and scientific knowledge. *Development and Change*, *26*(3), 413–439. https://doi.org/10.1111/j.1467-7660.1995.tb00560.x

Agrawal, A. (2008). The role of local institutions in adaptation to climate change. *Papers of the Social Dimensions of Climate Change Workshop*. The World Bank, Washington DC, March 5–6, 2008. https://openknowledge.worldbank.org/bitstream/handle/10986/28274/691280WP0P11290utions0in0adaptation.pdf?sequence=1&isAllowed=y

Agrawal, A., Kononen, M., & Perrin, N. (2009). *The role of local institutions in adaptation to climate change*. Papers of the Social Dimensions of Climate Change Workshop Social Development Department, The World Bank, Washington, DC, March 5–6, 2008. (Paper No. 118) Washington, DC: The World Bank.

Agrawal, A., & Perrin, N. (2008). *Climate adaptation, local institutions, and rural livelihoods*. IFRI Working Paper, W081-6. International Forestry Resources and Institutions Program, University of Michigan. http://websites.umich.edu/~ifri/Publications/W08I6%20Arun%20Agrawal%20and%20Nicolas%20Perrin.pdf

Agrawal, A., & Perrin, N. (2009). Climate adaptation, local institutions and rural livelihoods. In W. N. Adger, I. Lorenzoni, & K. L. O'Brien (Eds.), *Adapting to climate change: Thresholds, values, governance* (pp. 350–367). Cambridge: Cambridge University Press.

Agrawal, A., McSweeney, C., & Perrin, N. (2008). *Local institutions and climate change adaptation*. Social Development Notes No: 113. Washington, DC: World Bank. https://openknowledge.worldbank.org/bitstream/handle/10986/11145/4489100BRI0Box31te0Change0Adaptation.pdf?sequence=1&isAllowed=y

Ahmed, I., Maund, K., & Gajendran, T. (2020). *Disaster resilience in South Asia: Tackling the odds in the sub-continental fringes*. London: Routledge.

Ahmed, M., & Suphachalasai, S. (2014). *Assessing the costs of climate change and adaptation in South Asia*. Asian Development Bank. http://hdl.handle.net/11540/46

Aldrich, D. P., Page-Tan, C. M., & Paul, C. J. (2016). Social capital and climate change adaptation. In *Oxford research encyclopedia of climate science*. Retrieved 10 Nov. 2022, from https://oxfordre.com/climatescience/view/10.1093/acrefore/9780190228620.001.0001/acrefore-9780190228620-e-342

Aldrich, D., Page-Tan, C., & Paul, C. (n.d.) Social Capital and Climate Change Adaptation. *Oxford Research Encyclopedia of Climate Science.*

Aldrich, D. P. (2011). The externalities of strong social capital: Post-tsunami recovery in Southeast India. *Journal of Civil Society, 7*(1), 81–99. https://doi.org/10.1080/17448689.2011.553441

Alexander, C., Bynum, N., Johnson, E., King, U., Mustonen, T., Neofotis, P., et al. (2011). Linking indigenous and scientific knowledge of climate change. *BioScience, 61*(6), 477–484. https://doi.org/10.1525/bio.2011.61.6.10.

Allison, E., & Ellis, F. (2001). The livelihoods approach and management of small-scale fisheries. *Marine Policy, 25*(5), 377–388. https://doi.org/10.1016/S0308-597X(01)00023-9

Allison, E., & Horemans, B. (2006). Putting the principles of the sustainable livelihoods approach into fisheries development policy and practice. *Marine Policy, 30*(6), 757–766. https://doi.org/10.1016/j.marpol.2006.02.001

Allison, E. H., Adger, W. N., Badjeck, M. -C., Brown, K., Conway, D., Dulvy, N. K., Halls, A., Perry, A., & Reynolds, J. D. (2005). *Effects of climate change on the sustainability of capture and enhancement fisheries important to the poor: Analysis of the vulnerability and adaptability of fisherfolk living in poverty.* Project No. R4778J. Final Technical Report. https://assets.publishing.service.gov.uk/media/57a08ca340f0b652dd00145a/R4778Ja.pdf

Allison, E. H., Kurien, J., Ota, Y., et al. (2020). *The human relationship with our ocean planet.* Washington, DC: World Resources Institute. https://oceanpanel.org/blue-papers/HumanRelationshipwithOurOceanPlanet

Alsop, R., & Heinsohn, N. (2012). *Measuring empowerment in practice: Structuring analysis and framing indicators* (Vol. 3510). Washington, DC: World Bank.

Alsop, R., Bertelsen, M. F., & Holland, J. (2006). *Empowerment in practice: From analysis to implementation.* Washington, DC: World Bank.

Amrith, S. S. (2013, October 13). The Bay of Bengal: In Peril from climate change. *New York Times.* https://www.nytimes.com/2013/10/14/opinion/the-bay-of-bengal-in-peril-from-climate-change.html

Amrith, S. S. (2018). Risk and the South Asian monsoon. *Climatic Change, 151*(1), 17–28. https://doi.org/10.1007/s10584-016-1629-x

Amundsen, H. (2012). Illusions of resilience? An analysis of community responses to change in Northern Norway. *Ecology and Society, 17*(4). https://www.jstor.org/stable/26269221

Ananth Pur, K. (2007a). Dynamics of local governance in Karnataka. *Economic and Political Weekly, 42*(8), 667–673.

Ananth Pur, K. (2007b). Rivalry or synergy? Formal and informal local governance in rural India. *Development and Change, 38*(3), 401–421. https://doi.org/10.1111/j.1467-7660.2007.00417.x

Ananth Pur, K., & Moore, M. P. (2010). Ambiguous institutions: Traditional governance and local democracy in rural South India. *Journal of Development Studies, 46*(4), 603–623. https://doi.org/10.1080/00220380903002921

Anderson, J., & Bausch, C. (2006). *Climate change and natural disasters: Scientific evidence of a possible relation between recent natural disasters and climate change.* Briefing Note, Environment, Public Health and Food Safety Committee, European Parliament.

Andersson, K., & Agrawal, A. (2011). Inequalities, institutions, and forest commons. *Global Environmental Change, 21*(3), 866–875. https://doi.org/10.1016/j.gloenvcha.2011.03.004

Arivudai Nambi, A., & Sekhar Bahinipati, C. (2013). Adaptation to climate change and livelihoods: An integrated case study to assess the vulnerability and adaptation options of the fishing and farming communities of selected east coast stretch of Tamil Nadu, India. *Asian Journal of Environment and Disaster Management (AJEDM) – Focusing on Pro-Active Risk Reduction in Asia*, 4(3), 297–321. https://doi.org/10.3850/s1793924012001691

Ariyabandu, M. M., & Wickramasinghe, M. (2003). *Gender Dimensions in Disaster Management: A Guide for South Asia*. Warwickshire: Practical Action Publishing Ltd.

Arlikatti, S., & Andrew, S. A. (2012). Housing design and long-term recovery processes in the aftermath of the 2004 Indian Ocean tsunami. *Natural Hazards Review*, 13(1), 34–44. https://doi.org/10.1061/(ASCE)NH.1527-6996.0000062

Armitage, D., Berkes, F., Dale, A., Kocho-Schellenberg, E., & Patton, E. (2011). Co-management and the co-production of knowledge: Learning to adapt in Canada's Arctic. *Global Environmental Change*, 21(3), 995–1004. https://doi.org/10.1016/j.gloenvcha.2011.04.006.

Armitage, D., Charles, A., & Berkes, F. (Eds.). (2017). *Governing the coastal commons: Communities, resilience and transformation*. Abingdon: Routledge.

Aswani, S., Vaccaro, I., Abernethy, K., Albert, S., & Fernandez-Lopez de Pablo, J. (2015). Can perceptions of environmental and climate change in island communities assist in adaptation planning locally? *Environmental Management*, 56(6), 1487–1501. https://doi.org/10.1007/s00267-015-0572-3

Athiyaman, N. (2011). Nautical terms as gleaned from ancient tamil literature. In *Proceedings of the Asia Pacific Conference on Underwater Archaeology, Held at Manila, Philippines* (pp. 119–130). Conference Organising Committee, Manila.

Austin, Z., & Sutton, J. (2014). Qualitative research: Getting started. *The Canadian Journal of Hospital Pharmacy*, 67(6), 436–440. https://doi.org/10.4212%2Fcjhp.v67i6.1406

Ayeb-Karlsson, S., Tanner, T., van der Geest, K., Warner, K., Adams, H., Ahmedi, I., et al. (2015). *Livelihood resilience in a changing world – 6 global policy recommendations for a more sustainable future*. UNU-EHS Working Paper 22. UNU-EHS, Bonn.

Ayers, J., & Forsyth, T. (2009). Community-based adaptation to climate change. *Environment: Science and Policy for Sustainable Development*, 51(4), 22–31. https://doi.org/10.3200/ENV.51.4.22-31

Babu, S. (2011). Coastal accumulation in Tamil Nadu. *Economic and Political Weekly*, 46(48), 12–13.

Badjeck, M. C. (2008). *Vulnerability of coastal fishing communities to climate variability and change: Implications for fisheries livelihoods and management in Peru* (Doctoral dissertation). Universität Bremen. https://d-nb.info/989897052/34

Badjeck, M. C., Mendo, J., Wolff, M., & Lange, H. (2009). Climate variability and the Peruvian scallop fishery: The role of formal institutions in resilience building. *Climatic Change*, 94(1), 211–232. https://doi.org/10.1007/s10584-009-9545-y

Badjeck, M. C., Allison, E. H., Halls, A. S., & Dulvy, N. K. (2010). Impacts of climate variability and change on fishery-based livelihoods. *Marine Policy*, 34(3), 375–383. https://doi.org/10.1016/j.marpol.2009.08.007

Badola, E. (2020). *Fishing ban in Tamil Nadu: The need for a revision based on realities*. Policy Corps.

Baiju, K. K., Parappurathu, S., Abhilash, S., Ramachandran, C., Lekshmi, P. S., Padmajan, P., Padua, S., & Kaleekal, T. (2022). Achieving governance synergies through institutional interactions among non-state and state actors in small-scale marine fisheries in India. *Marine Policy*, 138, 104990. https://doi.org/10.1016/j.marpol.2022.104990

Bakker, K., del Moral, L., Downing, T., Giansante, C., Garrido, A., Iglesias, E., et al. (1999). *Societal and institutional responses to climate change and climatic hazards: Managing changing flood and drought risk* (SIRCH), K. Bakker (Ed.), SIRCH, Working Paper No. 3. https://citeseerx.ist.psu.edu/viewdoc/download?doi=10.1.1.462.2983&rep=rep1&type=pdf

Bal, P. K., Ramachandran, A., Palanivelu, K., Thirumurugan, P., Geetha, R., & Bhaskaran, B. (2016). Climate change projections over India by a downscaling approach using PRECIS. *Asia-Pacific Journal of Atmospheric Sciences*, 52(4), 353–369. https://doi.org/10.1007/s13143-016-0004-1

Balakrishnan, B. (2015). *Sanga elakkiyathil neiyethal nilapanpum nagarikamum* (In Tamil) (Doctoral dissertation). Manonmaniam Sundaranar University. http://hdl.handle.net/10603/179322

Bardhan, P. (2002). Decentralization of governance and development. *Journal of Economic Perspectives*, 16(4), 185–206. https://doi.org/10.1257/089533002320951037

Bavinck, M. (1996). Fisher regulations along the Coromandel coast: A case of collective control of common-pool resources. *Marine Policy*, 20(6), 475–482. https://doi.org/10.1016/S0308-597X(96)00039-5

Bavinck, M. (1998). "A matter of maintaining peace" state accommodation to subordinate legal systems: The case of fisheries along the Coromandel Coast of Tamil Nadu, India. *The Journal of Legal Pluralism and Unofficial Law*, 30(40), 151–170. https://doi.org/10.1080/07329113.1998.10756501

Bavinck, M. (2006). Women in fisheries. Don't be hasty and impetuous. Notes on women in fisheries. *Samudra*, 43, 3–5. https://www.icsf.net/wp-content/uploads/2021/06/43_Sam43_eng_all.pdf

Bavinck, M. (2001a). Caste panchayats and the regulation of fisheries along Tamil Nadu's Coromandel Coast. *Economic and Political Weekly*, 36(13), 1088–1094.

Bavinck, M. (2001b). *Marine resource management: Conflict and regulation in the fisheries of the Coromandel Coast*. New Delhi: Sage.

Bavinck, M. (2011). Wealth, poverty, and immigration: The role of institutions in the fisheries of Tamil Nadu, India. In S. Jentoft, & A. Eide (Eds.), *Poverty mosaics: Realities and prospects in small-scale fisheries* (pp. 173–191). Dordrecht: Springer.

Bavinck, M. (2016). The role of informal fisher village councils (ur panchayat) in Nagapattinam District and Karaikal, India. In S. Siar, & D. Kalikoski (Eds.), *Strengthening organizations and collective action in fisheries: Towards the formulation of a capacity development programme* (Vol. 41, pp. 383–404). (FAO Fisheries and Aquaculture Proceedings; Vol. 41). Rome: FAO.

Bavinck, M. (2020). Implications of legal pluralism for socio-technical transition studies–scrutinizing the ascendancy of the ring seine fishery in India. *The Journal of Legal Pluralism and Unofficial Law*, 52(2), 134–153. https://doi.org/10.1080/07329113.2020.1796297

Bavinck, M., & Jyotishi, A. (Eds.). (2014). *Conflict, negotiations and natural resource management: A legal pluralism perspective from India*. Abingdon and New York: Routledge.

Bavinck, M., & Salagrama, V. (2008). Assessing the governability of capture fisheries in the Bay of Bengal. *The Journal of Transdisciplinary Environmental Studies*, 7(1), 1–13. https://hdl.handle.net/11245/1.293274

Bavinck, M., & Vivekanandan, V. (2017). Qualities of self-governance and wellbeing in the fishing communities of northern Tamil Nadu, India-the role of Pattinavar ur panchayats. *Maritime Studies*, 16(1), 1–19. https://doi.org/10.1186/s40152-017-0070-8

Bebbington, A. (1999). Capitals and capabilities: A framework for analyzing peasant viability, rural livelihoods and poverty. *World Development, 27*(12), 2021–2044. https://doi.org/10.1016/S0305-750X(99)00104-7

Becken, S., Lama, A., & Espiner, S. (2013). The cultural context of climate change impacts: Perceptions among community members in the Annapurna Conservation Area, Nepal. *Environmental Development, 8*, 22–37. https://doi.org/10.1016/j.envdev.2013.05.007

Beermann, J., Damodaran, A., Jörgensen, K., & Schreurs, M. A. (2016). Climate action in Indian cities: An emerging new research area. *Journal of Integrative Environmental Sciences, 13*(1), 55–66. https://doi.org/10.1080/1943815X.2015.1130723

Berg, B. L. (2009). *Qualitative research methods for the social sciences* (7th ed.). Harlow: Pearson Education.

Berkes, F. (2008). *Sacred ecology* (1st ed.). New York: Routledge.

Berkes, F., Folke, C., & Colding, J. (Eds.). (2000). *Linking social and ecological systems: Management practices and social mechanisms for building resilience*. Cambridge: Cambridge University Press.

Berkes, F., Mahon, R., McConney, P., Pollnac, R., & Pomeroy, R. (2001). *Managing small-scale fisheries – Alternative directions and methods*. Ottawa: International Development Research Centre.

Berman, R. J. (2014). *Developing climate change coping capacity into adaptive capacity in Uganda* (Doctoral dissertation). University of Leeds. http://etheses.whiterose.ac.uk/7104/

Berman, R., Quinn, C., & Paavola, J. (2012). The role of institutions in the transformation of coping capacity to sustainable adaptive capacity. *Environmental Development, 2*, 86–100. https://doi.org/10.1016/j.envdev.2012.03.017

Bhagirathan, U., Meenakumari, B., Jayalakshmy, K. V., Panda, S. K., Madhu, V. R., & Vaghela, D. T. (2010). Impact of bottom trawling on sediment characteristics – A study along inshore waters off Veraval coast, India. *Environmental Monitoring and Assessment, 160*(1), 355–369. https://doi.org/10.1007/s10661-008-0700-0

Bhalla, R. S., Ram, S., & Srinivas, V. (2008). *Studies on vulnerability and habitat restoration along the Coromandel Coast* (245p). India: Publication of UNDP/UNTRS & FERAL.

Bharamappanavara, S. C. (2013). Growth and outreach of self-help groups' microcredit models in India: A literature insight. *International Journal of Social and Economic Research, 3*, 1–14. https://doi.org/10.5958/j.2249-6270.3.1.001

Bharathi, S. B. (1999). *Coromandel fishermen: An ethnography of pattanavar subcaste*. Pondicherry, India: Pondicherry Institute of Linguistics and Culture.

Bhathal, B. (2014). *Government-led development of India's marine fisheries since 1950: Catch and effort trends, and bioeconomic models for exploring alternative policies* (Doctoral dissertation). University of British Columbia. https://bit.ly/2RwjHTF

Bhattamishra, R., & Barrett, C. B. (2008). Community-based risk management arrangements: An overview and implications for social fund program design. https://ssrn.com/abstract=1141878 or http://dx.doi.org/10.2139/ssrn.1141878

Bhattamishra, R., & Barrett, C. (2010). Community-based risk management arrangements: A €review. *World Development, 38*(7), 923–932. https://doi.org/10.1016/j.worlddev.2009.12.017

Bjurström, A., & Polk, M. (2011). Physical and economic bias in climate change research: A scientometric study of IPCC third assessment report. *Climatic Change, 108*(1), 1–22. https://doi.org/10.1007/s10584-011-0018-8

Blaikie, P., Cannon, T., Davis, I., & Wisner, B. (1994). *At risk: Natural hazards, people's vulnerability and disasters*. London and New York: Routledge.

BMTPC. (2006). *An introduction to the vulnerability atlas of India – First revision*. Building Materials and Technology Promotion Council. https://www.bmtpc.org/admin/PublisherAttachement/An%20Introduction%20to%20the%20Vulnerability%20Atlas%20of%20IndiaAtt.pdf

Bogardi, J., & Warner, K. (2009). Here comes the flood. *Nature Climate Change, 1*(901), 9–11. https://doi.org/10.1038/climate.2008.138

Bouwer, L. M. (2011). Have disaster losses increased due to anthropogenic climate change? *Bulletin of the American Meteorological Society*. 6, 791–793. https://doi.org/10.1175/2010BAMS3092.1

Branisa, B., Klasen, S., & Ziegler, M. (2010) Why we should all care about social institutions related to gender inequality? *Proceedings of the German Development Economics Conference*, Hannover 2010, No. 50, Verein für Socialpolitik, Ausschuss für Entwicklungsländer, Göttingen.

Branisa, B., Klasen, S., & Ziegler, M. (2013). Gender inequality in social institutions and gendered development outcomes. *World Development, 45*, 252–268. https://doi.org/10.1016/j.worlddev.2012.12.003

Brenkert, A. L., & Malone, E. L. (2005). Modeling vulnerability and resilience to climate change: A case study of India and Indian states. *Climatic Change, 72*(1), 57–102. https://doi.org/10.1007/s10584-005-5930-3

Brierley, A. S., & Kingsford, M. J. (2009). Impacts of climate change on marine organisms and ecosystems. *Current Biology, 19*(14), R602–R614. https://doi.org/10.1016/j.cub.2009.05.046

Brink, H. I. (1993). Validity and reliability in qualitative research. *Curationis, 16*(2), 35–38. https://doi.org/10.4102/curationis.v16i2.1396

Brown, H. C. P., & Sonwa, D. J. (2015). Rural local institutions and climate change adaptation in forest communities in Cameroon. *Ecology and Society, 20*(2). http://dx.doi.org/10.5751/ES-07327-200206

Brulle, R. J., & Dunlap, R. E. (2015). Sociology and global climate change: Introduction. In R. E. Dunlap., & R. J. Brulle (Eds.), *Climate change and society sociological perspectives* (pp. 1–31). New York: Oxford University Press.

Bryant, R. L. (1992). Political ecology: An emerging research agenda in Third-World studies. *Political Geography, 11*(1), 12–36. https://doi.org/10.1016/0962-6298(92)90017-N

Bryld, E. (2001). Increasing participation in democratic institutions through decentralization: Empowering women and scheduled castes and tribes through panchayat raj in rural India. *Democratization, 8*(3), 149–172. https://doi.org/10.1080/714000213

Buckingham, S., & Masson, V. (Eds.). (2017). *Understanding climate change through gender relations*. London: Routledge.

Burt, R. (2000). The network structure of social capital. *Research In Organizational Behavior, 22*, 345–423. https://doi.org/10.1016/S0191-3085(00)22009-1

Busby, C. J. (1995). *Gender, person and exchange in a fishing community in Kerala, South India* (Doctoral dissertation). London School of Economics and Political Science, United Kingdom. https://core.ac.uk/download/pdf/46519213.pdf

Byg, A., & Salick, J. (2009). Local perspectives on a global phenomenon – Climate change in Eastern Tibetan villages. *Global Environmental Change, 19*(2), 156–166. https://doi.org/10.1016/j.gloenvcha.2009.01.010.

Byravan, S., Chellarajan, S., & Rangarajan, R. (2010). *Sea level rise: Impact on major infrastructure, ecosystems and land along the Tamil Nadu coast*. Chennai: Centre for Development Finance (CDF), IFMR and Humanities and Social Sciences, IIT Madras.

References

Byrne, M. M. (2001). Linking philosophy, methodology, and methods in qualitative research. *AORN Journal*, 73(1), 207–210. https://doi.org/10.1016/s0001-2092(06)62088-7

Cardona, O. D., van Aalst, M. K., Birkmann, J., Fordham, M., McGregor, G., Perez, R., Pulwarty, R. S., Schipper, E. L. F., & Sinh, B. T. (2012). Determinants of risk: Exposure and vulnerability. In C. B. Field, V. Barros, T. F. Stocker, D. Qin, D. J. Dokken, K. L. Ebi, M. D. Mastrandrea, K. J. Mach, G.-K. Plattner, S. K. Allen, M. Tignor, & P. M. Midgley (Eds.), *Managing the risks of extreme events and disasters to advance climate change adaptation. A special report of Working Groups I and II of the Intergovernmental Panel on Climate Change (IPCC)* (pp. 65–108). Cambridge, UK, and New York, NY, USA: Cambridge University Press.

CCC&AR and TNSCCC (2015). Climate Change Projection (Rainfall) for Nagapattinam. In: *District-Wise Climate Change Information for the State of Tamil Nadu*. Centre for Climate Change and Adaptation Research (CCC&AR), Anna University and Tamil Nadu State Climate Change Cell (TNSCCC), Department of Environment (DoE), Government of Tamil Nadu, Chennai, Tamil Nadu, India. Available at URL. www.tnsccc.in

Chakraborty, A., & Joshi, P. (2016). Mapping disaster vulnerability in India using analytical hierarchy process. *Geomatics, Natural Hazards and Risk*, 7(1), 308–325. http://dx.doi.org/10.1080/19475705.2014.897656

Chambers, R. (1989). Editorial introduction: Vulnerability, coping and policy. *IDS Bulletin*, 20(2), 1–7. https://doi.org/10.1111/j.1759-5436.1989.mp20002001.x

Chambers, R., & Conway, G. (1992). *Sustainable rural livelihoods: Practical concepts for the 21st century*. Institute of Development Studies (UK). https://opendocs.ids.ac.uk/opendocs/bitstream/handle/20.500.12413/775/Dp296.pdf?sequence=1&isAllowed=y

Chandrasekhar, D. (2010). *Understanding stakeholder participation in post-disaster recovery (case study: Nagapattinam, India* (Doctoral Dissertation). University of Illinois at Urbana-Champaign. http://hdl.handle.net/2142/16018

Charles, A. (2012). People, oceans and scale: Governance, livelihoods and climate change adaptation in marine social–ecological systems. *Current Opinion in Environmental Sustainability*, 4(3), 351–357. https://doi.org/10.1016/j.cosust.2012.05.011

Charrad, M. M. (2010, November). Women's agency across cultures: Conceptualizing strengths and boundaries. *Women's Studies International Forum*, 33(6), 517–522. https://doi.org/10.1016/j.wsif.2010.09.004

Chaudhary, D. R., Ghosh, A., & Patolia, J. S. (2006). Characterization of soils in the tsunami-affected coastal areas of Tamil Nadu for agronomic rehabilitation. *Current Science*, 91(1), 99–104.

Chishakwe, N., Murray, L., & Chambwera, M. (2012). *Building climate change adaptation on community experiences: Lessons from community-based natural resource management in southern Africa*. International Institute for Environment and Development. https://pubs.iied.org/sites/default/files/pdfs/migrate/10030IIED.pdf

Christensen, A.-S. (2007). *Methodological framework for studying fishermen's tactics and strategies*. Institute for Fisheries Management, Aalborg University. https://vbn.aau.dk/ws/portalfiles/portal/13048859/Workingpaper182007.pdf

Chuenpagdee, R., & Jentoft, S. (2007). Step zero for fisheries co-management: What precedes implementation. *Marine Policy*, 31(6), 657–668. https://doi.org/10.1016/j.marpol.2007.03.013

Chuenpagdee, R., & Song, A. M. (2012). Institutional thinking in fisheries governance: Broadening perspectives. *Current Opinion in Environmental Sustainability*, 4(3), 309–315. https://doi.org/10.1016/j.cosust.2012.05.006

Chuenpagdee, R., & Jentoft, S. (2015). Rethinking small-scale fisheries governance. In H. D. Smith, J. L. Suárez de Vivero, & T. S. Agardy (Eds.), *Routledge handbook of ocean resources and management* (pp. 241–254). London: Routledge.

Chuenpagdee, R., & Jentoft, S. (2018). Transforming the governance of small-scale fisheries. *Maritime Studies*, *17*(1), 101–115. https://doi.org/10.1007/s40152-018-0087-7

Chuenpagdee, R. (2019). Too big to ignore – A transdisciplinary journey. In R. Chuenpagdee, & S. Jentoft (Eds.), *Transdisciplinarity for small-scale fisheries governance* (pp. 15–31). Cham: Springer.

Cinner, J., Adger, W. N., Allison, E. H., Barnes, M. L., Brown, K., Cohen, P. J., Gelcich, S., Hicks, C. C., Hughes, T. P., Lau, J., Marshall, N. A., & Morrison, T. H. (2018). Building adaptive capacity to climate change in tropical coastal communities. *Nature Climate Change*, *8*(2), 117–123. https://doi.org/10.1038/s41558-017-0065-x

Clark, G. E., Moser, S. C., Ratick, S. J., Dow, K., Meyer, W. B., Emani, S., Jin, W., Kasperson, J. X., Kasperson, R. E., & Schwarz, H. E. (1998). Assessing the vulnerability of coastal communities to extreme storms: The case of Revere, MA., USA. *Mitigation and Adaptation Strategies for Global Change*, *3*(1), 59–82. https://doi.org/10.1023/A:1009609710795

Clark, B., & Clausen, R. (2008). The oceanic crisis: Capitalism and the degradation of marine ecosystems. *Monthly Review*, *60*(3), 91. https://doi.org/10.14452/MR-060-03-2008-07_6

CMFRI. (2012a). *Marine fisheries census (2010) Part 1. India*. New Delhi: Government of India, Ministry of Agriculture, Department of Animal Husbandry, Dairying & Fisheries and Central Marine Fisheries Research Institute, Indian Council of Agricultural Research. http://eprints.cmfri.org.in/8998/1/India_report_full.pdf

CMFRI. (2012b). *Marine fisheries census 2010 Part II. 4 Tamil Nadu*. New Delhi: Government of India, Ministry of Agriculture, Department of Animal Husbandry, Dairying & Fisheries and Central Marine Fisheries Research Institute, Indian Council of Agricultural Research. http://eprints.cmfri.org.in/9002/1/TN_report_full.pdf

Cochrane, P. (2006). Exploring cultural capital and its importance in sustainable development. *Ecological Economics*, *57*(2), 318–330. https://doi.org/10.1016/j.ecolecon.2005.04.012

Coleman, J. S. (1987). Norms as social capital. In G. Radni, & P. Bernholz (Eds.), *Economic imperialism: The economic approach applied outside the field of economics* (pp. 135–155). New York: Paragon House Publishers.

Colwell, J. M. N. (2016). *Fishery-dependent stakeholders-Impacts and responses to an annual closed fishing season in Tamil Nadu & Puducherry, India* (Doctoral dissertation). Michigan State University.

Colwell, J. M. N., Axelrod, M., & Roth, B. (2019). Unintended consequences of a seasonal ban on fishing effort in Tamil Nadu & Puducherry, India. *Fisheries Research*, *212*, 72–80. https://doi.org/10.1016/j.fishres.2018.12.003

Convention on Biological Diversity. (2016). *Sustainable fisheries*. CBD Press Brief. Canada: Secretariat of the Convention on Biological Diversity. https://www.cbd.int/idb/image/2016/promotional-material/idb-2016-press-brief-fish.pdf

Corbin, J., & Strauss, A. (1998). *Basics of qualitative research: Techniques and procedures for developing grounded theory* (2nd ed.). Thousand Oaks: Sage.

Cornea, N., Véron, R., & Zimmer, A. (2017). Clean city politics: An urban political ecology of solid waste in West Bengal, India. *Environment and Planning A*, *49*(4), 728–744. https://doi.org/10.1177%2F0308518X16682028

Coulthard, S. (2008). Adapting to environmental change in artisanal fisheries – Insights from a South Indian Lagoon. *Global Environmental Change*, *18*(3), 479–489. https://doi.org/10.1016/j.gloenvcha.2008.04.003

Crane, T. A. (2013). *The role of local institutions in adaptive processes to climate variability: The cases of southern Ethiopia and southern Mali*. Oxfam Research Report. https://oi-files-d8-prod.s3.eu-west-2.amazonaws.com/s3fs-public/file_attachments/rr-local-institutions-adaptive-climate-ethiopia-mali-080213-en_0.pdf

Creswell, J. W. (2014). *Research design: Qualitative, quantitative, and mixed methods approaches* (4th edition). Thousand Oaks, CA: Sage Publications.

Crouch, M., & McKenzie, H. (2006). The logic of small samples in interview-based qualitative research. *Social Science Information*, *45*(4), 483–499. https://doi.org/10.1177%2F0539018406069584

CSTEP. (2014). *Climate change vulnerability and adaptation in South Asia: Project report*. The Center for Study of Science, Technology and Policy (CSTEP). https://cstep.in/drupal/sites/default/files/2019-01/CSTEP_RR_Climate_Change_Vulnerability_and_Adaptation_in_South_Asia_2014.pdf

Cutter, S. (1995). The forgotten casualties: Women, children, and environmental change. *Global Environmental Change*, *5*(3), 181–194. https://doi.org/10.1016/0959-3780(95)00046-Q

Cutter, S. (1996). Vulnerability to environmental hazards. *Progress in Human Geography*, *20*(4), 529–539. https://doi.org/10.1177%2F030913259602000407

Cutter, S., & Finch, C. (2008). Temporal and spatial changes in social vulnerability to natural hazards. *Proceedings of the National Academy of Sciences*, *105*(7), 2301–2306. https://doi.org/10.1073/pnas.0710375105

Cutter, S. B., Osman-Elasha, J. Campbell, S.-M. Cheong, S. McCormick, R. Pulwarty, S. Supratid, & G. Ziervogel (2012). Managing the risks from climate extremes at the local level. In C. B. Field, V. Barros, T. F. Stocker, D. Qin, D. J. Dokken, K. L. Ebi, M. D. Mastrandrea, K. J. Mach, G.-K. Plattner, S. K. Allen, M. Tignor, & P. M. Midgley (Eds.), *Managing the risks of extreme events and disasters to advance climate change adaptation* (pp. 291–338). A Special Report of Working Groups I and II of the Intergovernmental Panel on Climate Change (IPCC). Cambridge, UK, and New York, NY, USA: Cambridge University Press.

Cutter, S., Boruff, B., & Shirley, W. (2003). Social vulnerability to environmental hazards. *Social Science Quarterly*, *84*(2), 242–261. https://doi.org/10.1111/1540-6237.8402002

d-maps. (n.d.). *Map of India – Boundaries, states*. https://d-maps.com/carte.php?num_car=4182&lang=en

Das, S. (2016). Television is more effective in bringing behavioral change: Evidence from heat-wave awareness campaign in India. *World Development*, *88*, 107–121. https://doi.org/10.1016/j.worlddev.2016.07.009

Das, S., & Smith, S. C. (2012). Awareness as an adaptation strategy for reducing mortality from heat waves: Evidence from a disaster risk management program in India. *Climate Change Economics*, *3*(02), 1250010. https://doi.org/10.1142/S2010007812500108

Deb, A. K. and Haque, C. E. (2016). Livelihood diversification as a climate change coping strategy adopted by small-scale fishers of Bangladesh. In. W. L. Filho., H. Musa., G. Cavan., P. O'Hare, & J. Seixas (Eds.), *Climate change adaptation, resilience and hazards* (pp. 345–368). Cham: Springer Publishers International.

Denton, F. (2002). Climate change vulnerability, impacts, and adaptation: Why does gender matter? *Gender & Development*, *10*(2), 10–20. https://doi.org/10.1080/13552070215903

Department of Fisheries and Fishermen Welfare. (n.d.). *Welfare schemes*. Department of Fisheries and Fishermen Welfare, Government of Tamil Nadu. https://www.fisheries.tn.gov.in/WelfareSchemes

Dessai, S., & Hulme, M. (2004). Does climate adaptation policy need probabilities? *Climate Policy*, *4*(2), 107–128. https://doi.org/10.1080/14693062.2004.9685515

DevendraBabu, M., Rajasekhar, D., & Manjula, R. (July–December, 2011). Role of Grama Sabha in the implementation of MGNREGS: Field insights from Karnataka, *The Grassroots Governance Journal*, *9*, 2.

DFID. (2001). *Sustainable livelihoods guidance sheets*. London: Department for International Development (DFID). https://www.livelihoodscentre.org/

Diaz-Cayeros, A., Magaloni, B., & Ruiz-Euler, A. (2014). Traditional governance, citizen engagement, and local public goods: Evidence from Mexico. *World Development*, *53*, 80–93. https://doi.org/10.1016/j.worlddev.2013.01.008

Divakarannair, N (2007). *Livelihood assets and survival strategies in coastal communities in Kerala, India* (Doctoral dissertation). University of Victoria. http://hdl.handle.net/1828/260

Donnelly-Roark, P., Ouedraogo, K., & Ye, X. (2001). *Can local institutions reduce poverty? Rural decentralization in Burkina Faso*. World Bank Policy Research Working Paper No. 2677, Environment and Social Development Unit, Africa Region, World Bank. http://hdl.handle.net/10986/19551

Doshi, S. (2017) Embodied urban political ecology: Five propositions. *Area*, *49*(1), 125–8. https://doi.org/10.1111/area.12293

Downing, T. E., & Patwardhan, A. (2004). Assessing vulnerability for climate adaptation. In B. Lim, & E. Spanger-Siegfried (Eds.), *Adaptation policy frameworks for climate change: Developing strategies, policies, and measures* (pp. 67–90). Cambridge: Cambridge University Press.

Drewes, E. (1986). *Activating fisherwomen for development through trained link workers in Tamil Nadu, India*. Madras: Development of Small-Scale Fisheries in the Bay of Bengal.

Dubash, N. K. (2009). Climate change and development: A bottom-up approach to mitigation for developing countries? In R. B. Stewart, B. Kingsbury, & B. Rudyk (Eds.), *Climate finance* (pp. 172–178). New York & London: New York University Press.

Duyne Barenstein, J. (2009). Housing reconstruction in Tamil Nadu. The disaster after the tsunami. In D. Miller., & J. Rivera (Eds.), *Community disaster recovery and resiliency: Exploring global opportunities and challenges* (pp. 343–362). Boca Raton: Auerbach Publications.

Duyne Barenstein, J. (2016). The right to adequate housing in post-disaster situations: The case of relocated communities in Tamil Nadu, India. In P. Daly, & R. M. Feener. (Eds.), *Rebuilding Asia following natural disasters: Approaches to reconstruction in the Asia-Pacific region* (pp. 236–260). Cambridge: Cambridge University Press.

Eguavoen, I., Schulz, K., De Wit, S., Weisser, F., & Müller-Mahn, D. (2015). Political dimensions of climate change adaptation: Conceptual reflections and African examples. In W. L. Filho (Ed.), *Handbook of climate change adaptation* (pp. 1183–1199). Berlin/Heidelberg: Springer.

Eisenhardt, K. M. (1989). Building theories from case study research. *Academy of Management Review*, *14*(4), 532–550. https://doi.org/10.5465/amr.1989.4308385

Ellis, F. (2000). *Rural livelihoods and diversity in developing countries*. Oxford: Oxford university press.

Enarson, E., & Chakrabarti, P. D. (Eds.). (2009). *Women, gender and disaster: Global issues and initiatives*. New Delhi: SAGE Publications India.

Enarson, E., & Fordham, M. (2001). From women's needs to women's rights in disasters. *Global Environmental Change Part B: Environmental Hazards*, *3*(3), 133–136. https://doi.org/10.3763/ehaz.2001.0314

Enarson, E., & Pease, B. (2016). *Men, masculinities and disaster*. London: Routledge.

ENVIS report. (n.d.). *Nagapattinam district*. ENVIS Centre: Tamil Nadu State of Environment and Related Issues. http://tnenvis.nic.in/files/NAGAPATTINAM%20%20.pdf

Eriksen, S. H., Nightingale, A. J., & Eakin, H. (2015). Reframing adaptation: The political nature of climate change adaptation. *Global Environmental Change*, *35*, 523–533. https://doi.org/10.1016/j.gloenvcha.2015.09.014

Eriksen, S., & Lind, J. (2009). Adaptation as a political process: Adjusting to drought and conflict in Kenya's drylands. *Environmental Management*, *43*(5), 817–835. https://doi.org/10.1007/s00267-008-9189-0

Faist, T. (2018). The socio-natural question: How climate change adds to social inequalities. *Journal of Intercultural Studies*, *39*(2), 195–206. https://doi.org/10.1080/07256868.2018.1446670

FAO. (1999). Guidelines for the routine collection of capture fishery data, FAO Fisheries Technical Paper 382. Prepared at the *FAO/DANIDA Expert Consultation, Bangkok, Thailand*, 18–30 May 1998, FAO, Rome, 1999. https://www.fao.org/3/x2465e/x2465e0h.htm#ANNEX%205.%20GLOSSARY

FAO. (2017). Towards gender-equitable small-scale fisheries governance and development – A handbook. *In support of the implementation of the Voluntary Guidelines for Securing Sustainable Small-Scale Fisheries in the Context of Food Security and Poverty Eradication*, by Nilanjana Biswas. Rome, Italy. https://www.fao.org/3/i7419e/i7419e.pdf

FAO. (2022a). *The state of world fisheries and aquaculture 2022. Towards blue transformation*. Rome: FAO. https://doi.org/10.4060/cc0461en

FAO. (2022b). *Fishing gear types. Purse seines. Technology fact sheets. Fisheries and Aquaculture Division [online]*. Rome. https://www.fao.org/fishery/en/geartype/249/en

Farrington, J. (2001). *Sustainable livelihoods, rights and the new architecture of aid. Natural Resource Perspectives* (Vol. 69). London, UK: Overseas Development Institute.

Farrington, J., Carney, D., Ashley, C., & Turton, C. (1999). *Sustainable livelihoods in practice: Early applications of concepts in rural areas* (Vol. 42, pp. 1–2). London: ODI.

Fernandez, A. (2006). History and spread of the self-help affinity group movement in India. *Occasional Paper*, *3*(6). International Fund for Agricultural Development (IFAD). https://myrada.org/wp-content/uploads/pdf/history3_sag_ifad.pdf

Field, E., Jayachandran, S., & Pande, R. (2010). Do traditional institutions constrain female entrepreneurship? A field experiment on business training in India. *American Economic Review*, *100*(2), 125–129. https://doi.org/10.1257/aer.100.2.125

Fihl, E. (2009). Introduction: The study of cultural encounters in Tharangampadi/Tranquebar. *Indo-Danish Cultural Encounters in Tranquebar: Past and Present*. Special issue of *Review of Development and Change*, 14(1–2), 7–18. http://natmus.dk/fileadmin/user_upload/natmus/forskning/dokumenter/Tranquebar/RDC_XIV_Tranquebar.pdf

Fihl, E. (2010). (Personal Communication). *Head of Panchyat in Nagipattinam, Participant observations, and Tranquebar panchayat cases*. (Raw data).

Fihl, E. (2013). The legitimacy of South Indian caste councils. In E. Fihl, & J. Dahl. (Eds.), *A comparative ethnography of alternative spaces* (pp. 41–63). New York: Palgrave Macmillan.

Fihl, E. (2014). The 'Second Tsunami': Disputed moralities of economic transactions among fishers. In: E. Fihl, & A. R. Venkatachalapathy (Eds.), *Beyond tranquebar: Grappling across cultural borders in South India* (pp. 123–147). Hyderabad: Orient Blackswan.

FIMSUL. (2011). *Fisheries management options for Tamil Nadu & Puducherry*. A Report prepared for the Fisheries Management for Sustainable Livelihoods (FIMSUL) Project, undertaken by the UN FAO in association with the World Bank, the Government of Tamil Nadu and the Government of Puducherry. Report No. FIMSUL/R20.FAO/UTF/IND/180/IND. New Delhi, Chennai and Puducherry, India: Vivekanandan, V., & Kasim, H. M.

Ford, J. D., Cameron, L., Rubis, J., Maillet, M., Nakashima, D., Willox, A. C., et al. (2016). Includ- ing indigenous knowledge and experience in IPCC assessment reports. *Nature Climate Change, 6*, 349–353. https://doi.org/10.1038/nclimate2954

Ford, J., & King, D. (2015). A framework for examining adaptation readiness. *Mitigation & Adaptation Strategies for Global Change, 20*(4), 505–526. https://doi.org/10.1007/s11027-013-9505-8

Ford, J., Keskitalo, E., Smith, T., Pearce, T., Berrang-Ford, L., Duerden, F., & Smit, B. (2010). Case study and analogue methodologies in climate change vulnerability research. *Wiley Interdisciplinary Reviews: Climate Change, 1*(3), 374–392. https://doi.org/10.1002/wcc.48

Fordham, M. (1999). The intersection of gender and social class in disaster: Balancing resilience and vulnerability. *International Journal of Mass Emergencies and Disasters, 17*(1), 15–36. https://doi.org/10.13140/2.1.1943.3601

Frangoudes, K., & Gerrard, S. (2018). (En) Gendering change in small-scale fisheries and fishing communities in a globalized world. *Maritime Studies, 17*(2), 117–124. https://doi.org/10.1007/s40152-018-0113-9

Füssel, H. M., & Klein, R. J. T. (2006). Climate change vulnerability assessments: An evolution of conceptual thinking. *Climatic Change, 75*(3), 301–329. https://doi.org/10.1007/s10584-006-0329-3

GADM (The Database of Global Administrative Areas). (n.d.). *GADM maps and data*. https://gadm.org/

Gaillard, J. C. (2010). Vulnerability, capacity and resilience: Perspectives for climate and development policy. *Journal of International Development: The Journal of the Development Studies Association, 22*(2), 218–232. https://doi.org/10.1002/jid.1675

Gandhi, K. (2011). *Identity, politics, power and resistance in disaster response: A case study of the 2004 tsunami, relief, rehabilitation and recovery in Tamil Nadu* (Doctoral dissertation). National University of Singapore, Singapore. http://scholarbank.nus.edu.sg/handle/10635/22061

Gangadharan, L., Jain, T., Maitra, P., & Vecci, J. (2018). The fairer sex? Women leaders and the strategic response to the social environment. *SSRN Electronic Journal*. https://dx.doi.org/10.2139/ssrn.2736033

Gayathri Devi, K. G. (2006). *Decentralization, governance and women heads of grama panchayats in Karnataka*. Report prepared for Centre for Decentralization and Development, Institute for Social and Economic Change, Bangalore

Geertz, A. (2004). *Seat reservation for women in local panchayats: An analysis of power* (Undergraduate Dissertation). South Asian Studies Hamilton College at New York, USA.

Geetha, R., Kizhakudan, S. J., Divipala, Indira, Salim, S. S., & Zacharia, P. U. (2017). Vulnerability index and climate change: An analysis in Cuddalore District of Tamil Nadu, India. *Indian Journal of Fisheries*, *64*(2), 96–104.

Gentle, P., Thwaites, R., Race, D., & Alexander, K. (2013). Changing role of local institutions to enable individual and collective actions for adapting to climate change. *IASC2013 Abstracts: The 14th Global Conference of the International Association for the Study of the Commons*, IASC, Japan. https://ro.uow.edu.au/cgi/viewcontent.cgi?article=1973&context=smhpapers

Ghosh, S., Vittal, H., Sharma, T., Karmakar, S., Kasiviswanathan, K. S., Dhanesh, Y., ... & Gunthe, S. S. (2016). Indian summer monsoon rainfall: Implications of contrasting trends in the spatial variability of means and extremes. *PloS One*, *11*(7), e0158670. https://doi.org/10.1371/journal.pone.0158670

Glantz, M. H. (1988). Politics and the air around us: International policy action on atmospheric pollution by trace gases. In M. H. Glantz (Ed.), *Societal responses to regional climatic change: Forecasting by analogy* (pp. 41–72). Boulder: Westview Press.

Glantz, M. H. (1995). Assessing the impacts of climate: The issue of winners and losers in a global climate change context. *Studies in Environmental Science*, *65*, 41–54. https://doi.org/10.1016/S0166-1116(06)80193-7

Glaser, B., & Strauss, A. (1967). Grounded theory: The discovery of grounded theory. *Sociology the Journal of the British Sociological Association*, *12*(1), 27–49.

Global Facility for Disaster Reduction and Recovery, GRDRR, World Bank. (2006). *Tsunami. India-Two Years After*. A joint report by the United Nations, the World Bank and the Asian Development Bank, GRDRR. https://www.gfdrr.org/sites/default/files/publication/TSunami%20_India_-Two_%20Years%20_After.pdf

Godfrey-Wood, R., & Flower, B. C. (2018). Does guaranteed employment promote resilience to climate change? The case of India's Mahatma Gandhi National Rural Employment Guarantee Act (MGNREGA). *Development Policy Review*, *36*, O586–O604. https://doi.org/10.1111/dpr.12309

Goh, A. H. X. (2012). *A literature review of the gender-differentiated impacts of climate change on women's and men's assets and well-being in developing countries*. CAPRi Working Paper No. 106. Washington, DC: International Food Policy Research Institute. http://dx.doi.org/10.2499/CAPRiWP106

Gomathy, N. B. (2006, January). The role of traditional panchayats in coastal fishing communities in Tamil Nadu, with special reference to their role in mediating tsunami relief and rehabilitation. In *Proceedings regional workshop on post-tsunami rehabilitation of fishing communities and fisheries-based livelihoods* (Vol. 18, p. 19).

Gomathy, N. B. (2011). *Coastal commons*. P2P Foundation Wiki. https://wiki.p2pfoundation.net/Coastal_Commons

Government of India. (2007). *Sixth report: Local governance. An inspiring journey into the future*. Second Administrative Reforms Commission. http://14.139.60.153/handle/123456789/2791

Government of India. (2017). *Mainstreaming disaster risk reduction & climate change adaptation in district level planning: A manual for district planning committees*. Disaster Management Division, Ministry of Home Affairs, Government of India & UNDP. https://ndmindia.mha.gov.in/images/pdf/DPCManual-final-printversion-lowresolution.pdf

Government of Tamil Nadu. (2004). *Tsunami relief the state of Tamil Nadu – India. Focus on Nagarpattinam experience*. Government of Tamil Nadu.

Government of Tamil Nadu. (2007). *India: Tsunami emergency assistance project (Tamil Nadu): Short resettlement plan for Kodiyampalayam bridge*. Prepared by highways department, Government of Tamil Nadu, India. https://www.adb.org/sites/default/files/project-document/78537/39114-06-ind-rp.pdf

Government of Tamil Nadu. (2013). *Tamil Nadu state action plan for climate change draft 2013*. Government of Tamil Nadu. https://cag.gov.in/uploads/media/tamil-nadu-climate-change-action-plan-20200726073516.pdf

Government of Tamil Nadu. (2018). *Fisheries department – Policy note, 2017–2018*. https://cms.tn.gov.in/sites/default/files/documents/fisheries_e_pn_2017_18.pdf

Government of Tamil Nadu. (2019). *Fisheries. Fisheries development 2018–19*. Commissioner of Fisheries, Director of the Marine Products Export Development Authority, Government of Tamil Nadu. https://www.tn.gov.in/deptst/fisheries.pdf

Government of Tamil Nadu. (2020). *Amendment to the Tamil Nadu marine fishing regulation rules, 1983* (Tamil Nadu Government Gazette, 17th February 2020). Animal Husbandry, Dairying and Fisheries Department, Government of Tamil Nadu. http://www.stationeryprinting.tn.gov.in/extraordinary/2020/65_Ex_III_1a.pdf

Government of Tamil Nadu. (2021). *Fisheries and fishermen welfare policy note 2021–2022*. Demand No. 7. Animal Husbandry, Dairying and Fisheries Department, Government of Tamil Nadu. https://cms.tn.gov.in/sites/default/files/documents/fisheries_e_pn_2021_22.pdf

Government of Tamil Nadu. (2022a). *Fisheries and fishermen welfare policy note 2022–2023*. Demand No. 7. Animal Husbandry, Dairying and Fisheries Department, Government of Tamil Nadu. https://cms.tn.gov.in/sites/default/files/documents/fisheries_e_pn_2022_23.pdf

Government of Tamil Nadu. (2022b). *Annexure*. Tamil Nadu Public Service Commission. https://www.tnpsc.gov.in/static_pdf/document/instructions-to-candiates.pdf

Grafton, R. (2005). Social capital and fisheries governance. *Ocean & Coastal Management, 48*(9–10), 753–766. https://doi.org/10.1016/j.ocecoaman.2005.08.003

Guerreiro, A., Ladle, R., & Batista, V. (2017). Erratum to: Riverine fishers' knowledge of extreme cli- matic events in the Brazilian Amazonia. *Journal of Ethnobiology and Ethnomedicine*. https://doi.org/10.1186/s13002-017-0143-1

Guhathakurta, P., & Revadekar, J. (2017). Observed variability and long-term trends of rainfall over India. In M. N. Rajeevan., & S. Nayak (Eds.), *Observed climate variability and change over the Indian region* (pp. 1–15). Singapore: Springer.

Guhathakurta, P., Sanap, S., Menon, P., Prasad, A. K., Sangwan, N., & Advani, S. C. (2020). *Observed rainfall variability and changes over Andhra Pradesh state*. Pune, India: India Meteorological Department. https://imdpune.gov.in/hydrology/rainfall%20variability%20page/tamil_final.pdf

Guhathakurta, P., Sreejith, O. P., & Menon, P. A. (2011). Impact of climate change on extreme rainfall events and flood risk in India. *Journal of Earth System Science, 120*(3), 359–373. https://doi.org/10.1007/s12040-011-0082-5

Gunakar, S., Jadhav, A., & Bhatta, R. (2017). Protections for small-scale fisheries in india: A study of india's monsoon fishing ban. In S. Jentoft., R. Chuenpagdee., M. J. Barragán-Paladines., & N. Franz. (Eds.), *The small-scale fisheries guidelines* (Vol. 14, pp. 291–311). MARE Publication Series (MARE). Cham: Springer.

Gupta, H. (2015) Women and climate change: Linking ground perspectives to the global scenario. *Indian Journal of Gender Studies, 22*(3), 408–420. https://doi.org/10.1177/0971521515594278

Gupta, J., Termeer, C., Klostermann, J., Meijerink, S., van den Brink, M., Jong, P., Nooteboom, S., & Bergsma, E. (2010). The adaptive capacity wheel: A method to assess the inherent characteristics of institutions to enable the adaptive capacity of society. *Environmental Science & Policy, 13*(6), 459–471. https://doi.org/10.1016/j.envsci.2010.05.006

Habtezion, S. (2016). *Gender and Climate change: Overview of linkages between gender and climate change*. New York. https://reliefweb.int/report/world/gender-and-climate-change-overview-linkages-between-gender-and-climate-change

Halder, P., Sharma, R., & Alam, A. (2012). Local perceptions of and responses to climate change: Experiences from the natural resource-dependent communities in India. *Regional Environmental Change, 12*, 665–673. https://doi.org/10.1007/s10113-012-0281-x

Halfdanardottir, J. (1993). Social mobilization in Kerala: Fishers, priests, unions, and political parties. *MAST. Maritime Anthropological Studies, 6*(1–2), 136–156.

Hanmer, L., & Klugman, J. (2016). Exploring women's agency and empowerment in developing countries: Where do we stand? *Feminist Economics, 22*(1), 237–263. https://doi.org/10.1080/13545701.2015.1091087

Hanson, J. L., Balmer, D. F., & Giardino, A. P. (2011). Qualitative research methods for medical educators. *Academic Pediatrics, 11*(5), 375–386. https://doi.org/10.1016/j.acap.2011.05.001

Hapke, H. M. (2012). capturing the complexities of globalization in fisheries: Gendered divisions of labour and divisions of labour and difference. *Asian Fisheries Science S, 25*, 75–92.

Hastrup, F. (2011). *Weathering the world: Recovery in the wake of the tsunami in a Tamil fishing village*. Oxford and New York: Berghahn Books.

Helmke, G., & Levitsky, S. (2004). Informal institutions and comparative politics: A research agenda. *Perspectives on Politics, 2*(4), 725–740. https://doi.org/10.1017/S1537592704040472

Herbert, V. (n.d.). *The sangam poems* (Vaidehi Herbert, Trans.). https://sangamtranslationsbyvaidehi.com/ettuthokai-akananuru-301-400/

Hines, R. I. (2007). Natural disasters and gender inequalities: The 2004 tsunami and the case of India. *Race, Gender & Class, 14*(1/2), 60–68. https://www.jstor.org/stable/41675195

Hiwasaki, L., Luna, E., Syamsidik, & Shaw, R. (2014). Process for integrating local and indigenous knowl- edge with science for hydro-meteorological disaster risk reduction and climate change adaptation in coastal and small island communities. *International Journal of Disaster Risk Reduction, 10*(Part A), 15–27. https://doi.org/10.1016/j.ijdrr.2014.07.007.

Hodgson, G. M. (2006). What are institutions? *Journal of Economic Issues, 40*(1), 1–25. https://doi.org/10.1080/00213624.2006.11506879

Hoegh-Guldberg, O., & Bruno, J. F. (2010). The impact of climate change on the world's marine ecosystems. *Science, 328*(5985), 1523–1528. https://doi.org/10.1126/science.1189930

Hoeppe, G. (2007). *Conversations on the beach. Fishermen's knowledge, metaphor and environmental change in South India*. New York & Oxford: Berghahn books.

Holling, C. (1973). Resilience and stability of ecological systems. *Annual Review of Ecology and Systematics, 4*(1), 1–23. https://doi.org/10.1146/annurev.es.04.110173.000245

Hoon, P., & Hyden, G. (2003). Governance and sustainable livelihood. In O. Bressers, & W. A. Rosenbaum. (Eds.), *Achieving sustainable development – The challenge of governance across social scales* (pp. 43–63). Westport: Praeger.

Hulme, M. (2009). *Why we disagree about climate change: Understanding controversy, inaction and opportunity*. Cambridge: Cambridge University Press.

References

Huq, S., & Reid, H. (2007). *Community based adaptation: A vital approach to the threat climate change poses to the poor*. London: IIED Briefing. https://hdl.handle.net/10535/6228

Hust, E. (2002). *Political Representation and Empowerment: Women in the Institutions of Local Government in Orissa after the 73rd Amendment to the Indian Constitution*. Heidelberg Papers in South Asian Comparative Politics. Working Paper No. 6. South Asia Institute, Department of Political Science, University of Heidelberg. https://doi.org/10.11588/heidok.00004098

ICOR. (2011). *Vulnerability to climate change to ecological and climate changes: A pilot study in Dharavi belt in Mumbai*. Mumbai: Institute for community organization research.

IFAD. (2005). *Post-tsunami sustainable livelihoods programme for the coastal communities of Tamil Nadu*. International Fund for Agricultural Development. https://webapps.ifad.org/members/eb/84/docs/EB-2005-84-R-16-Rev-2.pdf

IFAD. (2014). *Guidelines for integrating climate change adaptation into fisheries and aquaculture projects*. International Fund for Agricultural Development (IFAD). https://www.ifad.org/documents/38714170/39135645/fisheries.pdf/17225933-cea1-436d-a6d8-949025d78fbd

Immanuel, S., Pillai, V. N., Vivekanandan, E., Kurup, K. N., & Srinath, M. (2003). A preliminary assessment of the coastal fishery resources in India – Socio-economic and bioeconomic perspective. In G. Silvestre, L. Garces, I. Stobutzki, M. Ahmed, R. A. ValmonteSantos, C. Luna, L. Lachica Aliño, P. Munro, V. Christensen, & D. Pauly (Eds.), *Assessment, management and future directions for coastal fisheries in Asian countries* (Vol. 67, pp. 439–478). WorldFish Center Conference Proceedings. http://eprints.cmfri.org.in/8511/1/Chapter-21-FA.pdf

Indira, P., & Stephen, Rajkumar Inbanathan, S. (2013). Studies on the trend and chaotic behavior of Tamil Nadu rainfall. *Journal of Indian Geophysical Union, 17*(4), 335–339.

Infochange. (2010). *On the waterfront*. Centre for Communication and Development Studies. http://www.ccds.in/download/publication/agenda/agenda_18.pdf

Inglin, S. (2013). Links between post-tsunami relocation and changes in fi shing practices in Tamil Nadu: A micro-level case study. In J. E. Duyne Barenstein., & E. Leemann (Eds.), *Post-disaster reconstruction and change: A community perspective* (pp. 241–257). Boca Raton: CRC Press; Taylor & Francis.

IPCC. (2007). *Climate change 2007: Impacts, adaptation and vulnerability. Contribution of Working Group II to the fourth assessment report of the intergovernmental panel on climate change* [M. L. Parry., O. F. Canziani., J. P. Palutikof., P. J. van der Linden., & C. E. Hanson (Eds.)]. Cambridge University Press.

IPCC. (2012). *Managing the risks of extreme events and disasters to advance climate change adaptation. A special report of working groups I and II of the intergovernmental panel on climate change* [C. B. Field, V. Barros, T. F. Stocker, D. Qin, D. J. Dokken, K. L. Ebi, M. D. Mastrandrea, K. J. Mach, G.-K. Plattner, S. K. Allen, M. Tignor, & P. M. Midgley (Eds.)]. Cambridge University Press, Cambridge, UK, and New York, NY, USA, 582 pp.

IPCC. (2014a). *Climate change 2014: Synthesis report. Contribution of working groups I, II and III to the fifth assessment report of the intergovernmental panel on climate change*. IPCC, Geneva, Switzerland.

IPCC. (2014b). *Summary for policymakers. In: Climate change 2014: Impacts, adaptation, and vulnerability. Part A: Global and sectoral aspects. Contribution of Working Group II to the fifth assessment report of the intergovernmental panel on climate change* [C. B. Field, V. R. Barros, D. J. Dokken, K. J. Mach, M. D. Mastrandrea, T. E. Bilir, M. Chatterjee, K. L. Ebi, Y. O. Estrada, R. C. Genova, B. Girma, E. S. Kissel, A. N. Levy, S. MacCracken, P. R. Mastrandrea, & L. L. White (Eds.)]. Cambridge University Press, Cambridge, United Kingdom and New York, NY, USA, pp. 1–32.

IPCC. (2022). Annex II: Glossary [V. Möller, R. van Diemen, J. B. R. Matthews, C. Méndez, S. Semenov, J. S. Fuglestvedt, & A. Reisinger (Eds.)]. In *Climate change 2022: Impacts, adaptation and vulnerability. Contribution of working group II to the sixth assessment report of the intergovernmental panel on climate change* [H.-O. Pörtner, D. C. Roberts, M. Tignor, E. S. Poloczanska, K. Mintenbeck, A. Alegría, M. Craig, S. Langsdorf, S. Löschke, V. Möller, A. Okem, B. Rama (Eds.)]. Cambridge University Press, Cambridge, UK and New York, NY, USA, pp. 2897–2930, https://doi.org/10.1017/9781009325844.029.

Islam, M. M. (2013). *Vulnerability and adaptation of fishing communities to the impacts of climate variability and change: Insights from coastal Bangladesh.* University of Leeds. https://etheses.whiterose.ac.uk/5321/1/Islam%20MM_Earth%20&%20Environment_PhD%20Thesis%202013.pdf

Islam, M. M., & Herbeck, J. (2013). Migration and translocal livelihoods of coastal small-scale fishers in Bangladesh. *The Journal of Development Studies, 49*(6), 832–845. https://doi.org/10.1080/00220388.2013.766719

Islam, M. R., & Morgan, W. J. (2012). Non-governmental organizations in Bangladesh: Their contribution to social capital development and community empowerment. *Community Development Journal, 47*(3), 369–385. https://doi.org/10.1093/cdj/bsr024

Islam, M. M., Sallu, S., Hubacek, K., & Paavola, J. (2014a). Vulnerability of fishery-based livelihoods to the impacts of climate variability and change: Insights from coastal Bangladesh. *Regional Environmental Change, 14*(1), 281–294. https://doi.org/10.1007/s10113-013-0487-6

Islam, M. M., Sallu, S., Hubacek, K., & Paavola, J. (2014b). Limits and barriers to adaptation to climate variability and change in Bangladeshi coastal fishing communities. *Marine Policy, 43*, 208–216. https://doi.org/10.1016/j.marpol.2013.06.007

Islam, M. M., Sallu, S., Hubacek, K., & Paavola, J. (2014c). Migrating to tackle climate variability and change? Insights from coastal fishing communities in Bangladesh. *Climatic Change, 124*(4), 733–746. https://doi.org/10.1007/s10584-014-1135-y

Islam, R., & Walkerden, G. (2014). How bonding and bridging networks contribute to disaster resilience and recovery on the Bangladeshi coast. *International journal of disaster risk reduction, 10*, 281–291. https://doi.org/10.1016/j.ijdrr.2014.09.016

Islam, R., & Walkerden, G. (2015). How do links between households and NGOs promote disaster resilience and recovery? A case study of linking social networks on the Bangladeshi coast. *Natural hazards, 78*(3), 1707–1727. https://doi.org/10.1007/s11069-015-1797-4

Jacobson. C., Crevello, S., Chea, C., & Jarihani, B. (2019). When is migration a maladaptive response to climate change? *Regional Environmental Change, 19*(1), 101–112. https://doi.org/10.1007/s10113-018-1387-6

Jäger, J., Frühmann, J., Grünberger, S., & Vag, A. (2009). *Environmental change and forced migration scenarios, Synthesis Report.* Hungary: ATLAS Innoglobe Ltd.

Janakarajan, S. (2007). Challenges and prospects for adaptation: Climate and disaster risk reduction in coastal Tamilnadu. In M. Moench, & A. Dixit (Eds.), *Working with the winds of change* (pp. 235–270). Boulder, CO: ISET.

Janssen, M. A., Schoon, M. L., Ke, W., & Börner, K. (2006). Scholarly networks on resilience, vulnerability and adaptation within the human dimensions of global environmental change. *Global Environmental Change, 16*(3), 240–252. https://doi.org/10.1016/j.gloenvcha.2006.04.001

Jaswal, A. K., Rao, P. C. S., & Singh, V. (2015). Climatology and trends of summer high temperature days in India during 1969–2013. *Journal of Earth System Science, 124*(1), 1–15. https://doi.org/10.1007/s12040-014-0535-8

Jayaraman, T. (2015). Can political ecology comprehend climate change? *Review of Agrarian Studies, 4*(2). http://ras.org.in/can_political_ecology_comprehend_climate_change

Jentoft, S. (2004). Institutions in fisheries: What they are, what they do, and how they change. *Marine Policy, 28*(2), 137–149. https://doi.org/10.1016/S0308-597X(03)00085-X

Jentoft, S., Bavinck, M., Johnson, D., & Thomson, K. (2009). Fisheries co-management and legal pluralism: How an analytical problem becomes an institutional one. *Human Organization, 68*(1), 27–38. https://doi.org/10.17730/humo.68.1.h87q04245t63094r

Jentoft, S., & Chuenpagdee, R. (2009). Fisheries and coastal governance as a wicked problem. *Marine Policy, 33*(4), 553–560. https://doi.org/10.1016/j.marpol.2008.12.002

Johnson, D. S. (2018). The values of small-scale fisheries. In D. S. Johnson, T. G. Acott, N. Stacey, & J. Urquhart (Eds.), *Social wellbeing and the values of small-scale fisheries* (Vol 17, pp. 1–22). MARE Publication Series (MARE). Cham: Springer.

Jones, L. (2010). *Overcoming social barriers to adaptation*. Overseas Development Institute (ODI) Background Note. https://ssrn.com/abstract=2646812

Jordan, J. (2014). Swimming alone? The role of social capital in enhancing local resilience to climate stress: A case study from Bangladesh. *Climate and Development, 7*(2), 110–123. https://doi.org/10.1080/17565529.2014.934771

Jörgensen, K., Jogesh, A., & Mishra, A. (2015). Multi-level climate governance and the role of the subnational level. *Journal of Integrative Environmental Sciences, 12*(4), 235–245. https://doi.org/10.1080/1943815X.2015.1096797

Juran, L. (2012). The gendered nature of disasters: Women survivors in post-tsunami Tamil Nadu. *Indian Journal of Gender Studies, 19*(1), 1–29. https://doi.org/10.1177%2F097152151101900101

Kabeer, N. (1999). Resources, agency, achievements: Reflections on the measurement of women's empowerment. *Development and Change, 30*(3), 435–464. https://doi.org/10.1111/1467-7660.00125

Kabeer, N. (2008). *Paid work, women's empowerment and gender justice: Critical pathways of social change*. Pathways of Empowerment working papers (3). Institute of Development Studies, Brighton. http://eprints.lse.ac.uk/53077/1/Kabeer_Paid-work_Published.pdf

Kalikoski, D., QuevedoNeto, P., & Almudi, T. (2010). Building adaptive capacity to climate variability: The case of artisanal fisheries in the estuary of the Patos Lagoon, Brazil. *Marine Policy, 34*(4), 742–751. http://dx.doi.org/10.1016/j.marpol.2010.02.003

Kandlikar, M., & Sagar, A. (1999). Climate change research and analysis in India: An integrated assessment of a South-North divide. *Global Environmental Change, 9*, 119–138.

Karan, P. P., & Shanmugam, P. S. (2011). *The Indian Ocean Tsunami: The global response to a natural disaster*. Lexington: The University Press of Kentucky.

Karlsson, M., & Hovelsrud, G. K. (2015). Local collective action: Adaptation to coastal erosion in the Monkey River Village, Belize. *Global Environmental Change, 32*, 96–107. https://doi.org/10.1016/j.gloenvcha.2015.03.002

Keck, M., & Sakdapolrak, P. (2013). What is social resilience? Lessons learned and ways forward. *Erdkunde, 67*(1), 5–19. https://www.jstor.org/stable/23595352

Kelly, P. M., & Adger, W. N. (2000). Theory and practice in assessing vulnerability to climate change and facilitating adaptation. *Climatic Change*, *47*(4), 325–352. https://doi.org/10.1023/A:1005627828199

Kelman, I. (2006a). Warning for the 26 December 2004 tsunamis. *Disaster Prevention and Management*, *15*, 178–189.

Kelman, I. (2006b). Acting on disaster diplomacy. *Journal of International Affairs*, *59*(2), 215–240. https://www.jstor.org/stable/24358434

Kelman, I. (2010). Hearing local voices from small island developing states for climate change. *Local Environment*, *15*(7), 605–619. https://doi.org/10.1080/13549839.2010.498812

Kelman, I. (2011a). Understanding vulnerability to understand disasters. *Canadian disaster management textbook* (cap. 7). Canada: Canadian Risk and Hazards Network. https://citeseerx.ist.psu.edu/viewdoc/download?doi=10.1.1.517.3203&rep=rep1&type=pdf

Kelman, I. (2011b). *Disaster diplomacy: How disasters affect peace and conflict*. London: Routledge.

Kelman, I. (2014). No change from climate change: Vulnerability and small island developing states. *The Geographical Journal*, *180*(2), 120–129. https://doi.org/10.1111/geoj.12019

Kelman, I. (2016). Catastrophe and conflict: Disaster diplomacy and its foreign policy implications. *Brill Research Perspectives in Diplomacy and Foreign Policy*, *1*(1), 1–76. https://doi.org/10.1163/24056006-12340001

Kelman, I. (2020). *Disaster by choice: How our actions turn natural hazards into catastrophes*. Oxford: Oxford University Press.

Kelman, I., & West, J. J. (2009). Climate change and small island developing states: A critical review. *Ecological and Environmental Anthropology*, *5*(1), 1–16.

Kelman, I., Gaillard, J. C., & Mercer, J. (2015). Climate change's role in disaster risk reduction's future: Beyond vulnerability and resilience. *International Journal of Disaster Risk Science*, *6*(1), 21–27. https://doi.org/10.1007/s13753-015-0038-5

Kelman, I., Gaillard, J. C., Lewis, J., & Mercer, J. (2016). Learning from the history of disaster vulnerability and resilience research and practice for climate change. *Natural Hazards*, *82*(1), 129–143. https://doi.org/10.1007/s11069-016-2294-0

Kelman, I., Mercer, J., & Gaillard, J. C. (2012). Indigenous knowledge and disaster risk reduction. *Geography*, *97*(1), 12–21. https://doi.org/10.1080/00167487.2012.12094332

Kelman, I., Mercer, J., & West, J. J. (2009). Combining different knowledges: Community-based climate change adaptation in small island developing states. In H. Reid (Ed.), *Community-based adaptation to climate change* (pp. 41–53). London: International Institute for Environment and Development. https://pubs.iied.org/sites/default/files/pdfs/migrate/G02812.pdf

Kelman, I., Spence, R., Palmer, J., Patel, M., & Saito, K. (2008). Tourists and disasters: Lessons from the 26 December 2004 tsunamis. *Journal of Coastal Conservation*, *12*, 105–113. https://doi.org/10.1007/s11852-008-0029-4

Kenney, S. J. (1996). New research on gendered political institutions. *Political Research Quarterly*, *49*(2), 445–466. https://doi.org/10.1177%2F106591299604900211

Keys, N., Thomsen, D., & Smith, T. (2014). Adaptive capacity and climate change: The role of community opinion leaders. *Local Environment*, *21*(4), 432–450. https://doi.org/10.1080/13549839.2014.967758

Khan, A. S., Ramachandran, A., Usha, N., Aram, I. A., & Selvam, V. (2012a). Rising sea and threatened mangroves: A case study on stakeholders, engagement in climate change communication and non-formal education. *International Journal of Sustainable Development & World Ecology*, 19(4), 330–338. https://doi.org/10.1080/13504509.2011.650230

Khan, A. S., Ramachandran, A., Usha, N., Punitha, S., & Selvam, V. (2012b). Predicted impact of the sea-level rise at Vellar–Coleroon estuarine region of Tamil Nadu coast in India: Mainstreaming adaptation as a coastal zone management option. *Ocean & Coastal Management*, 69, 327–339. https://doi.org/10.1016/j.ocecoaman.2012.08.005

Kizhakudan, S. J., Raja, S., Gupta, K., Vivekanandan, E., Kizhakudan, J. K., Sethi, S. N., & Geetha, R. (2014). Correlation between changes in sea surface temperature and fish catch along Tamil Nadu coast of India-an indication of impact of climate change on fisheries? *Indian Journal of Fisheries*, 61(3), 111–115.

Kodirekkala, K. R. (2018). Cultural adaptation to climate change among indigenous people of South India. *Climatic Change*, 147(1–2), 299–312. https://doi.org/10.1007/s10584-017-2116-8

Kollmair, M., & Gamper, S. 2002. *The Sustainable Livelihoods Approach*. Input Paper for the Integrated Training Course of NCCR North-South Aeschiried, Switzerland (9–20 September 2002). Zurich, Switzerland: Development Study Group, University of Zurich.

Korakandy, R. (2008). *Fisheries development in India – A political economy of unsustainable development*. Delhi: Kalpaz publications.

Koralagama, D., Gupta, J., & Pouw, N. (2017). Inclusive development from a gender perspective in small scale fisheries. *Current Opinion in Environmental Sustainability*, 24, 1–6. https://doi.org/10.1016/j.cosust.2016.09.002

Krishna, A. (2002). *Active social capital: Tracing the roots of development and democracy (Ed)*. New York: Columbia University Press.

Krüger, F., Bankoff, G., Cannon, T., Orlowski, B., & Schipper, E. L. F. (Eds.). (2015). *Cultures and disasters understanding cultural framings in disaster risk reduction*. Abingdon and New York: Routledge.

Kruks-Wisner, G. (2011). Seeking the local state: Gender, caste, and the pursuit of public services in post-tsunami India. *World Development*, 39(7), 1143–1154. https://doi.org/10.1016/j.worlddev.2010.11.001

Kudva, N., & Misra, K. (2008). Gender quotas, the politics of presence, and the feminist project: What does the Indian experience tell us? *Signs: Journal of Women in Culture and Society*, 34(1), 49–73. https://doi.org/10.1086/589239

Kumar, N. (2017). Incentives and expectations: Community resiliency and recovery in Tamil Nadu after the Indian Ocean Tsunami. *Independent Review*, 22(1), 135–151.

Kumar, V. S., Pathak, K. C., Pednekar, P. Raju, N. S. N., & Gowthaman, R. (2006). Coastal processes along the Indian coastline. *Current Science*, 91(4), 530–536. https://www.jstor.org/stable/24093957

Kumaraperumal, R., Natarajan, S., Sivasamy, R., Chellamuthu, S., Ganesh, S. S., & Anandakumar, G. (2007). Impact of tsunami 2004 in coastal villages of Nagapattinam district, India. *Science of Tsunami Hazards*, 26(2), 93.

Kume, T., Umetsu, C., & Palanisami, K. (2009). Impact of the December 2004 tsunami on soil, groundwater and vegetation in the Nagapattinam district, India. *Journal of environmental management*, 90(10), 3147–3154. https://doi.org/10.1016/j.jenvman.2009.05.027

Kuppuswamy, S., & Rajarathnam, S. (2009). Women, information technology and disaster management: Tsunami affected districts of Tamil Nadu. *International Journal of Innovation and Sustainable Development*, 4(2–3), 206–215.

Kurien, J. (1996). *Towards a new agenda for sustainable small-scale fisheries development* (p. 48). Thiruvananthapuram: South Indian Federation of Fishermen Societies. https://dlc.dlib.indiana.edu/dlc/bitstream/handle/10535/5118/Towards%20a%20new%20agenda%20for%20sustainable%20small%20scale%20fisheries.pdf?sequence=1

Kurien, J. (1998). *Small-scale fisheries in the context of globalisation*. Thiruvananthapuram: Centre for Development Studies. http://localhost:8080/xmlui/handle/123456789/302

Kurien, J. (2007). The blessing of the commons: Small-scale fisheries, community property rights, and coastal natural assets. In J. K. Boyce, & S. Narain (Eds.), *Reclaiming nature: Environmental justice and ecological restoration* (pp. 23–52). London & New York: Anthem Press.

Kurien, J., & Paul, A. (2001). Social security nets for marine fisheries: The growth and changing composition of social security programmes in the fisheries sector of Kerala State, India. Thiruvananthapuram: Centre for Development Studies. http://localhost:8080/xmlui/handle/123456789/215

Kurien, J. (2020). *Community property rights: Re-establishing them for a secure future for small-scale fisheries*. Use of Property Rights in Fisheries Management, December 2020, Perth, Australia. http://publications.azimpremjiuniversity.edu.in/id/eprint/2444

Kuruppu, N., & Willie, R. (2015). Barriers to reducing climate enhanced disaster risks in Least developed country-small Islands through anticipatory adaptation. *Weather and Climate Extremes*, 7, 72–83. https://doi.org/10.1016/j.wace.2014.06.001

Lakshmi, A., Purvaja, R., & Ramesh, R. (2014). Recovering from the 2004 Indian Ocean tsunami: Lessons for climate change response. In B. C. Glavovic., & G. P. Smith (Eds.), *Adapting to climate change* (pp. 291–314). Dordrecht: Springer.

Larson, G., Drolet, J., & Samuel, M. (2015). The role of self-help groups in post-tsunami rehabilitation. *International Social Work*, 58(5), 732–742. https://doi.org/10.1177%2F0020872813477880

Lavell, A. M. Oppenheimer, Diop, C., Hess, J., Lempert, R., Li, J., Muir-Wood, R., & Myeong, S. (2012). Climate change: New dimensions in disaster risk, exposure, vulnerability, and resilience. In *Managing the risks of extreme events and disasters to advance climate change adaptation. A Special Report of Working Groups I and II of the Intergovernmental Panel on Climate Change (IPCC)* [C. B. Field, V. Barros, T. F. Stocker, D. Qin, D. J. Dokken, K. L. Ebi, M. D. Mastrandrea, K. J. Mach, G.-K. Plattner, S. K. Allen, M. Tignor, & P. M. Midgley (Eds.)]. Cambridge University Press, Cambridge, UK, and New York, NY, USA, pp. 25–64.

Le Comple, M. D., & Goetz, J. P. (1982). Problems of reliability and validity in ethnographic research. *Review of Educational Research*, 52(1), 31–60.

Lele, M. K. (2001). Local government: Conflict of interests and issues of legitimisation. *Economic and Political Weekly*, 4702–4704.

Li, L., Switzer, A. D., Wang, Y., Chan, C. H., Qiu, Q., & Weiss, R. (2018). A modest 0.5-m rise in sea level will double the tsunami hazard in Macau. *Science Advances*, 4(8), eaat1180. https://doi.org/10.1126/sciadv.aat1180

Longo, S. B., Clausen, R., & Clark, B. (2015). *The tragedy of the commodity: Oceans, fisheries, and aquaculture*. New Brunswick, New Jersey, and London: Rutgers University Press.

Longo, S. B., & Clark, B. (2016). An ocean of troubles: Advancing marine sociology. *Social Problems*, 63(4), 463–479. https://doi.org/10.1093/socpro/spw023

Lorber. (1994). *Paradoxes of gender*. New Haven, CT: Yale University.

Lorenz, D. F. (2010). The diversity of resilience: Contributions from a social science perspective. *Natural Hazards*, *67*(1), 7–24. https://doi.org/10.1007/s11069-010-9654-y

Lorenzoni, I., Nicholson-Cole, S., & Whitmarsh, L. (2007). Barriers perceived to engaging with climate change among the UK public and their policy implications. *Global Environmental Change*, *17*(3–4), 445–459. https://doi.org/10.1016/j.gloenvcha.2007.01.004

Macchi, M., Gurung, A. M., & Hoermann, B. (2015). Community perceptions and responses to climate variability and change in the Himalayas. *Climate and Development*, *7*(5), 414–425. https://doi.org/10.1080/17565529.2014.966046

MacDonald, R. (2005). How women were affected by the tsunami: A perspective from Oxfam. *PLoS Medicine*, *2*(6), e178. https://doi.org/10.1371/journal.pmed.0020178

Madhanagopal, D. (2018). Insecure lives under extreme climate conditions: Insights from a fishing hamlet in Tamil Nadu, India. *Metropolitics*. 17 April 2018. https://metropolitics.org/Insecure-Lives-Under-Extreme-Climate-Conditions-Insights-from-a-Fishing-Hamlet.html

Madhanagopal, D., & Pattanaik, S. (2020). Exploring fishermen's local knowledge and perceptions in the face of climate change: The case of coastal Tamil Nadu, India. *Environment, Development and Sustainability*, *22*, 3461–3489. https://doi.org/10.1007/s10668-019-00354-z

Madhanagopal, D., & Jacob, V. A. (2022). Climate risks in an unequal society: The question of climate justice in India. In D. Madhanagopal., C. T. Beer., B. R. Nikku., & A. Pelser (Eds.), *Environment, climate, and social justice* (pp. 161–188). Singapore: Springer.

Mahanayak, B., & Panigrahi, A. K. (2021). Sustainable management of fishermen co-operative societies in India: A review. *International Journal of Fisheries and Aquatic Studies*, *9*(3), 253–259.

Mahon, R., McConney, P., & Roy, R. (2008). Governing fisheries as complex adaptive systems. *Marine Policy*, *32*(1), 104–112. https://doi.org/10.1016/j.marpol.2007.04.011

Maiti, S., Jha, S. K., Garai, S., Nag, A., Chakravarty, R., Kadian, K.S. et al. (2015). Assessment of social vulnerability to climate change in the eastern coast of India. *Climatic Change*, *131*(2), 287–306. https://doi.org/10.1007/s10584-015-1379-1

Mandelbaum, D G (1970). *Society in India, 2 vols*, Berkeley: University of California Press.

Manor, J. (2003). *Democratic decentralisation in India*. Swedish International Development Cooperation Agency (Sida). https://cdn.sida.se/publications/files/sida2192en-democratic-decentralisation-in-india-2003-2007.pdf

Manuel-Navarrete, D. (2010). Power, realism, and the ideal of human emancipation in a climate of change. *Wiley Interdisciplinary Reviews: Climate Change*, *1*(6), 781–785. https://doi.org/10.1002/wcc.87

Mariappan, S., Felix, S., & Kalaiarasan, M. (2017). A present scenario of seine nets fishing gears of pulicat coast, Tamil Nadu, India. *International Journal of Fisheries and Aquatic Studies*, *5*(3), 208–212.

Marschke, M., Lykhim, O., & Kim, N. (2014). Can local institutions help sustain livelihoods in an era of fish declines and persistent environmental change? A Cambodian case study. *Sustainability*, *6*(5), 2490–2505. https://doi.org/10.3390/su6052490

Marshall, M. N. (1996). Sampling for qualitative research. *Family Practice*, *13*(6), 522–526. https://doi.org/10.1093/fampra/13.6.522

Marshall, N. A., Marshall, P. A., Tamelander, J., Obura, D., Malleret-King, D., & Cinner, J. E. (2010). *A framework for social adaptation to climate change: Sustaining tropical coastal communities [sic] and industries*. Gland, Switzerland: International Union for Conservation of Nature (IUCN). v + 36 pp.

Mascarenhas, A. (2004). Oceanographic validity of buffer zones for the east coast of India: A hydrometeorological perspective. *Current Science, 86*(3), 399–406.

Mascarenhas, A. (2006). Extreme events, intrinsic landforms and humankind: Post-tsunami scenario along Nagore–Velankanni coast, Tamil Nadu, India. *Current Science, 90*(9), 1195–1201.

Mathew, N. P. (2009) *Marine fisheries conservation and management in India*. United Nations – The Nippon Foundation Fellowship Programme. https://www.un.org/depts/los/nippon/unnff_programme_home/fellows_pages/fellows_papers/mathew_0809_india.pdf

Mathur, N. (2015). It's a conspiracy theory and climate change: Of beastly encounters and cervine disappearances in Himalayan India, *HAU: Journal of Ethnographic Theory, 5*(1), 87–111.

Mazzocchi, F. (2006). Western science and traditional knowledge: Despite their variations, different forms of knowledge can learn from each other. *EMBO Report. Science and Society Viewpoint, 7*(5), 463–466. https://doi.org/10.1038/sj.embor.7400693

McCarthy, J. J., Canziani, O. F., Leary, N. A., Dokken, D. J., & White, K. S. (Eds.). (2001). *Climate change 2001: Impacts, adaptation, vulnerability*. Cambridge, UK: Cambridge University Press.

McGranahan, G., Balk, D., & Anderson, B. (2007). The rising tide: Assessing the risks of climate change and human settlements in low elevation coastal zones. *Environment and Urbanization, 19*(1), 17–37. https://doi.org/10.1177%2F0956247807076960

McGuire, B. (2010). Potential for a hazardous geospheric response to projected future climate changes. *Philosophical Transactions of the Royal Society A: Mathematical, Physical and Engineering Sciences, 368*(1919), 2317–2345. https://doi.org/10.1098/rsta.2010.0080

McGuire, B. (2012). *Waking the giant – How a changing climate triggers earthquakes, tsunamis, and volcanoes*. New York: Oxford University Press.

McGuire, B., & Maslin, M. (2013). *Climate forcing of geological hazards*. Hoboken: Wiley.

McGuire, B., Mason, I., & Kilburn, C. (2002). *Natural hazards and environmental change*. London: Arnold.

McNamara, K. E., & Prasad, S. S. (2014). Coping with extreme weather: Communities in Fiji and Vanu- atu share their experiences and knowledge. *Climate Change, 123*, 121–132. https://doi.org/10.1007/s10584-013-1047-2

McNamara, K. E., & Westoby, R. (2011). Local knowledge and climate change adaptation on Erub Island, Torres Strait. *Local Environment, 16*(9), 887–901. https://doi.org/10.1080/13549839.2011.615304.

Mearns, R., & Norton, A. (2010). *Social dimensions of climate change: Equity and vulnerability in a warming world*. Washington: World Bank Publication.

Mercer, J. (2010). Disaster risk reduction or climate change adaptation: Are we reinventing the wheel? *Journal of International Development: The Journal of the Development Studies Association, 22*(2), 247–264. https://doi.org/10.1002/jid.1677

Mercer, J., Kelman, I., Taranis, L., & Suchet-Pearson, S. (2010). Framework for integrating indigenous and scientific knowledge for disaster risk reduction. *Disasters, 34*(1), 214–239. https://doi.org/10.1111/j.1467-7717.2009.01126.x

Miller, F., & McGregor, A. (2020). Rescaling political ecology? World regional approaches to climate change in the Asia Pacific. *Progress in Human Geography*, *4*(4), 663–682. https://doi.org/10.1177%2F0309132519849292

Mishra, N. K., & Tripathi, T. (2011). Conceptualising Women's agency, autonomy and empowerment. *Economic and Political Weekly*, *46*(11), 58–65.

Mistry, J., & Berardi, A. (2016). Bridging indigenous and scientific knowledge. *Science*, *352*(6291), 1274–1275. https://doi.org/10.1126/science.aaf1160.

Mitchell, T., & Tanner, T. M. (2006). *Adapting to climate change: Challenges and opportunities for the development community*. Teddington: Tearfund.

Mohmand, S. K. (2005). Representative decentralization vs. participatory decentralization: A critical analysis of local government plan 2000. In L. C. Jain (Ed.), *Decentralization and local governance*. New Delhi: Orient Longman.

Momtaz, S., & Shameem, M. (2015). *Experiencing climate change in Bangladesh vulnerability and adaptation in coastal regions*. London: Academic Press.

Moosa, C., & Tuana, N. (2014). Mapping a research agenda concerning gender and climate change: A review of the literature. *Hypatia*, *29*(3), 677–694. https://doi.org/10.1111/hypa.12085

Morse, J. (2001). Situating grounded theory within qualitative inquiry. In R. Schreiber., & P. N. Stern (Eds.), *Using grounded theory in nursing* (pp. 1–16). New York: Springer.

Moser, C., & Norton, A. (2001). *To claim our rights. Livelihood security, human rights and sustainable development*. London: ODI.

Moser, S. C., & Ekstrom, J. A. (2010). A framework to diagnose barriers to climate change adaptation. *Proceedings of the National Academy of Sciences*, *107*(51), 22026–22031. https://doi.org/10.1073/pnas.1007887107

Mubaya, C. P., & Mafongoya, P. (2017). The role of institutions in managing local level climate change adaptation in semi-arid Zimbabwe. *Climate Risk Management*, *16*, 93–105. https://doi.org/10.1016/j.crm.2017.03.003

Mukherjee, S., Aadhar, S., Stone, D., & Mishra, V. (2018). Increase in extreme precipitation events under anthropogenic warming in India. *Weather And Climate Extremes*, *20*, 45–53. https://doi.org/10.1016/j.wace.2018.03.005

Murty, T. S., Aswathanarayana, U., & Nirupama, N. (Eds.). (2007). *The Indian Ocean tsunami*. London: CRC Press.

Murugan, A., & Durgekar, R. (2008). *Beyond the Tsunami: Status of fisheries in Tamil Nadu, India: A snapshot of present and long-term trends*. Bangalore, India: UNDP/UNTRS, Chennai and ATREE. Accessed 06 January 2018 from https://www.dakshin.org/wp-content/uploads/2013/04/FISHERIES-REPORT.pdf.

Musinguzi, L., Efitre, J., Odongkara, K., Ogutu-Ohwayo, R., Muyodi, F., Natugonza, V., Olokotum, M., Namboowa, S., & Naigaga, S. (2016). Fishers' perceptions of climate change, impacts on their livelihoods and adaptation strategies in environmental change hotspots: A case of Lake Wamala, Uganda. *Environment, Development and Sustainability*, *18*(4), 1255–1273. https://doi.org/10.1007/s10668-015-9690-6

Nagapattinam District Disaster Management Report. (2012). *Nagapattinam District Disaster Management Report*. Government of Tamil Nadu (Not available online).

Nagy, G., Gómez-Erache, M., & Kay, R. (2015). A risk-based and participatory approach to assessing climate vulnerability and improving governance in coastal Uruguay. In B. Glavovic, M. Kelly, R. Kay, & A. Travers (Eds.), *Climate change and the coast building resilient communities* (1st ed., pp. 357–377). London: Taylor and Francis – CRC Press.

Nakashima, D., & Roue, M. (2002). Indigenous knowledge, peoples and sustainable practice. In P. Timmerman (Ed.), *Encyclopedia of global environmental change* (pp. 314–324). Chichester: Wiley.

Namboothri, N., Subramanian, D., Muthuraman, B., Sridhar, A., Rodriguez, S., & Shanker, K. (2008). Beyond the tsunami: Coastal sand dunes of Tamil Nadu, India – An overview. *UNDP/UNTRS, Chennai and ATREE, Bangalore, India.* https://www.dakshin.org/wp-content/uploads/2018/08/SAND-DUNES-REPORT.pdf

Nandakumar, D., & Muralikrishna, M. (1998). Mapping the extent of coastal regulation zone violations of the Indian coast. In *Report for National Fish workers Forum, Valiathura, Thiruvananthapuram.*

Narasimhan, C. V. (2004). *The Tamil language: A brief history of the language and its literature.* Chennai: Indian Institute of Technology Madras.

Natesan, U., Parthasarathy, A., Vishnunath, R., Kumar, G. E. J., & Ferrer, V. A. (2015). Monitoring longterm shoreline changes along Tamil Nadu, India using geospatial techniques. *Aquatic Procedia, 4*, 325–332. https://doi.org/10.1016/j.aqpro.2015.02.044

Nath, P. K., & Behera, B. (2011). A critical review of impact of and adaptation to climate change in developed and developing economies. *Environment, Development and Sustainability, 13*(1), 141–162. https://doi.org/10.1007/s10668-010-9253-9

National Remote Sensing Center (ISRO, Government of India). *Bhuvan.* Hyderabad, India: Indian Geo Platform of ISRO. https://bhuvan.nrsc.gov.in/home/index.php

Nayak, N. (2016). *In conversation: Nalini Nayak.* Sahapedia. Retrieved from https://www.sahapedia.org/conversation-nalini-nayak

Nayak, P. K. (2015). Institutional Pluralism, Multilevel Arrangements and Polycentricism: The Case of Chilika Lagoon, India. In M. Bavinck, & A. Jyotishi (Eds.), *Conflict, negotiations and natural resource management: A legal pluralism perspective from India* (pp. 148–177). London: Routledge

Nayak, N. (2017a). *Conversation: Nalini Nayak.* Sahapedia. https://www.sahapedia.org/conversation-nalini-nayak

Nayak, P. K. (2017b). Fisher communities in transition: Understanding change from a livelihood perspective in Chilika Lagoon, India. *Maritime Studies, 16*(1), 1–33. https://doi.org/10.1186/s40152-017-0067-3

NCAER. (2010). *Impact assessment and economic benefits of weather and marine services.* New Delhi: National Council of Applied Economic Research. https://www.moes.gov.in/sites/default/files/2021-07/ImpactAssessment-MOES.pdf

NCW. (n.d.). *Effectiveness of women self-help groups in micro enterprise development in Rajasthan and Tamil Nadu.* National Commission for Women, Government of India. http://ncw.nic.in/content/effectiveness-women-self-help-groups-micro-enterprise-development-rajasthan-and-tamil-nadu

NDMA. (2019). Study report on Gaja Cyclone 2018. National Disaster Management Authority (NDMA), Ministry of Home Affairs, Government of India. https://tnsdma.tn.gov.in/app/webroot/img/document/publications/gaja_cyclone_status_report.pdf

NDTV. (2018a, November 30). *Cyclone Gaja shatters dreams in Tamil Nadu, brings back Tsunami memories.* NDTV. https://www.ndtv.com/india-news/cyclone-gaja-shatters-dreams-in-tamil-nadu-brings-back-tsunami-memories-1955587

NDTV (2018b, November 28). *E. Palaniswami visits Cyclone Gaja-hit areas amid demands for more help.* NDTV. https://www.ndtv.com/tamil-nadu-news/tamil-nadu-chief-minister-e-palaniswami-to-visit-cyclone-gaja-hit-areas-amid-demands-for-more-help-1954468

Nehren, U., Subedi, J., Yanakieva, I., Sandholz, S., Pokharel, J., Chandra Lal, A., et al. (2013). Community perception on climate change and climate-related disaster preparedness in Kathmandu Valley, Nepal. *Journal of Natural Resources and Development, 4*, 35–57. https://doi.org/10.5027/jnrd.v3i0.04

Nelson, D. R., Adger, W. N., & Brown, K. (2007). Adaptation to environmental change: Contributions of a resilience framework. *Annual review of Environment and Resources,32*(1), 395–419. https://doi.org/10.1146/annurev.energy.32.051807.090348

Nielsen, J. Ø., & Reenberg, A. (2010). Cultural barriers to climate change adaptation: A case study from Northern Burkina Faso. *Global Environmental Change, 20*(1), 142–152. https://doi.org/10.1016/j.gloenvcha.2009.10.002

Nordgren, J., Stults, M., & Meerow, S. (2016). Supporting local climate change adaptation: Where we are and where we need to go. *Environmental Science & Policy, 66*, 344–352. https://doi.org/10.1016/j.envsci.2016.05.006

North, D. C. (1991). Institutions. *Journal of Economic Perspectives, 5*(1), 97–112. https://doi.org/10.1257/jep.5.1.97

Novak, J. M., & Axelrod, M. (2016). Patterns of multi-level fisheries governance and their impact on fishermen's adaptation strategies in Tamil Nadu, India. *Environmental Policy and Governance, 26*(1), 45–58. https://doi.org/10.1002/eet.1694

O'Brien, K. L., & Leichenko, R. M. (2000). Double exposure: Assessing the impacts of climate change within the context of economic globalization. *Global Environmental Change, 10*(3), 221–232. https://doi.org/10.1016/S0959-3780(00)00021-2

Odero, K. (2003). *Extending sustainable livelihoods framework*. Harare, Zimbabwe: Department of Rural and Urban Planning, University of Zimbabwe. https://hdl.handle.net/10535/4221

OECD. (2010). *The economics of adapting fisheries to climate change*. OECD Publishing. http://dx.doi.org/10.1787/9789264090415-en

Olsson, L., Opondo, M., Tschakert, P., Agrawal, A., Eriksen, S. H., Ma, S., Perch, L. N., & Zakieldeen, S. A. (2014). Livelihoods and poverty. In C. B. Field, V. R. Barros, D. J. Dokken, K. J. Mach, M. D. Mastrandrea, T. E. Bilir, M. Chatterjee, K. L. Ebi, Y. O. Estrada, R. C. Genova, B. Girma, E. S. Kissel, A. N. Levy, S. MacCracken, P. R. Mastrandrea, & L. L. White (Eds.), *Climate change 2014: Impacts, adaptation, and vulnerability. Part A: Global and sectoral aspects. Contribution of Working Group II to the fifth assessment report of the intergovernmental panel on climate change* (pp. 793-782). Cambridge University Press, Cambridge, United Kingdom and New York, NY, USA.

Onta, N., & Resurreccion, B. P. (2011). The role of gender and caste in climate adaptation strategies in Nepal. *Mountain Research and Development, 31*(4), 351–356. https://doi.org/10.1659/MRD-JOURNAL-D-10-00085.1

Orlove, B., Roncoli, C., Kabugo, M., & Majugu, A. (2010). Indigenous climate knowledge in southern Uganda: The multiple components of a dynamic regional system. *Climatic Change, 100*(2), 243–265. https://doi.org/10.1007/s10584-009-9586-2

Osbahr, H., Twyman, C., Adger, W., & Thomas, D. (2010). Evaluating successful livelihood adaptation to climate variability and change in Southern Africa. *Ecology and Society, 15*(2). https://doi.org/10.5751/es-03388-150227

Ostrom, E. (1995). *Governing the commons: The evolution of institutions for collective action*. New York: Cambridge University Press.

Otto, I. M., Reckien, D., Reyer, C. P. O., Marcus, R., Le Masson, V., & Jones, L., Norton, A., & Serdeczny, O. (2017). Social vulnerability to climate change: A review of concepts and evidence. *Regional Environmental Change, 17*(6), 1651–1662. https://doi.org/10.1007/s10113-017-1105-9

Oxfam. (2005). *The Tsunami's impact on women*. Oxfam briefing note. Oxfam International. https://oxfamilibrary.openrepository.com/bitstream/handle/10546/115038/bn-tsunami-impact-on-women-250305-en.pdf

Paerregaard, K. (2016). Making sense of climate change. Global impacts, local responses, and anthropogenic dilemmas. In S. A. Crate, & M. Nuttall (Eds.), *Anthropology and climate change* (pp. 251–260). New York: Routledge.

Paerregaard, K. (2018). The climate-development nexus: Using climate voices to prepare adaptation initiatives in the Peruvian Andes. *Climate and Development*, *10*(4), 360–368. https://doi.org/10.1080/17565529.2017.1291400

Pahl-Wostl, C. (2009). A conceptual framework for analysing adaptive capacity and multi-level learning processes in resource governance regimes. *Global Environmental Change*, *19*(3), 354–365. https://doi.org/10.1016/j.gloenvcha.2009.06.001

Palanithurai, G. (2001). Synergization of institutional arrangements for development and social justice at the grassroots. In S. P. Jain (Ed.), *A case analysis in emerging institutions for decentralized rural development* (Vol. II). Hyderabad: National Institute of Rural Development.

Palanithurai, G. (2008). *Critical issues in decentralization of powers in India*. New Delhi: Concept Publishing Company.

Palanithurai, G., & Ragupathy, V. (2008). *Communities, panchayats and governance at grassroots*. New Delhi: Concept Publishing Company.

Palinkas, L. A., Horwitz, S. M., Green, C. A., Wisdom, J. P., Duan, N., & Hoagwood, K. (2015). Purposeful sampling for qualitative data collection and analysis in mixed method implementation research. *Administration and Policy in Mental Health and Mental Health Services Research*, *42*(5), 533–544. https://doi.org/10.1007/s10488-013-0528-y

Panayotou, T. (1982). *Management concepts for small-scale fisheries: Economic and social aspects*. FAO Fisheries Technical Papers No. 228, Rome: FAO.

Panda, S. M. (2006). *Women's collective action and sustainable water management: Case of SEWA's water campaign in Gujarat, India*. CAPRi Working Paper No. 61. Washington, DC: International Food Policy Research Institute.

Panditharatne, C. (2016). Institutional barriers in adapting to climate change: A case study in Sri Lanka. *Ocean & Coastal Management*, *130*, 73–78. https://doi.org/10.1016/j.ocecoaman.2016.06.003

Panigrahi, J. K., & Mohanty, P. K. (2012). Effectiveness of the Indian coastal regulation zones provisions for coastal zone management and its evaluation using SWOT analysis. *Ocean & Coastal Management*, *65*, 34–50. https://doi.org/10.1016/j.ocecoaman.2012.04.023

Parthasarathy, D. (2015). Informality, resilience, and the political implications of disaster governance. *Pacific Affairs*, *88*(3), 551–575. https://doi.org/10.5509/2015883551

Parthasarathy, R., & Gupta, M. (2014). Mangrove management and cyclone risk reduction in Kachchh, Gujarat. In R. Shaw (Ed.), *Disaster recovery: Used or misused development opportunity* (pp. 301–315). Tokyo: Springer.

Pascual-Fernández, J. J. (1999). Participative management of artisanal fisheries in the Canary Islands. In D. Symes (Ed.), *Europe's Southern Waters: Management Issues and Practice* (pp. 66–77). Oxford: Fishing New Books.

Pascual-Fernández, J. J., Frangoudes, K., & Williams, S. (2005). Local institutions. In J. Kooiman, M. Bavinck, S. Jentoft, & R. Pullin (Eds.), *Fish for life: Interactive governance for fisheries* (pp. 153–172). Amsterdam: Amsterdam University Press.

Patnaik, U., & Narayanan, K. (2005). *Vulnerability and climate change: An analysis of the Eastern coastal districts of India.* International Workshop on Human Security and Climate Change. https://mpra.ub.uni-muenchen.de/22062/

Patnaik, U., & Narayanan, K. (2009). *Vulnerability and Climate Change: An Analysis of the Eastern Coastal Districts of India.* In Munich Personal RePEc Archive, Paper No: 22062.

Patt, A. G., & Schröter, D. (2008). Perceptions of climate risk in Mozambique: Implications for the success of adaptation strategies. *Global Environmental Change, 18*(3), 458–467. https://doi.org/10.1016/j.gloenvcha.2008.04.002

Patton, M. Q. (2002). *Qualitative research and evaluation methods* (3rd ed.). Thousand Oaks: Sage Publications.

Paul, C. J., Weinthal, E. S., Bellemare, M. F., & Jeuland, M. A. (2016). Social capital, trust, and adaptation to climate change: Evidence from rural Ethiopia. *Global Environmental Change, 36,* 124–138. https://doi.org/10.1016/j.gloenvcha.2015.12.003

Pauly, D. (2009, September 28). Apocalypse now. The New Republic. https://newrepublic.com/article/69712/aquacalypse-now

Pauly, D., Christensen, V., Guénette, S., Pitcher, T. J., Sumaila, U. R., Walters, C. J., Watson, R., & Zeller, D. (2002). Towards sustainability in world fisheries. *Nature, 418*(6898), 689–695. https://doi.org/10.1038/nature01017

Pearse, R. (2017). Gender and climate change. *Wiley Interdisciplinary Reviews: Climate Change, 8*(2), e451. https://doi.org/10.1002/wcc.451

Pelling, M. (2010). *Adaptation to climate change: From resilience to transformation.* London: Routledge.

Pelling, M., & High, C. (2005). Understanding adaptation: What can social capital offer assessments of adaptive capacity? *Global Environmental Change, 15*(4), 308–319. https://doi.org/10.1016/j.gloenvcha.2005.02.001

Pelling, M., & Manuel-Navarrete, D. (2011). From resilience to transformation: The adaptive cycle in two Mexican urban centers. *Ecology and Society, 16*(2). https://www.jstor.org/stable/26268885

Perry, R., Ommer, R., Barange, M., & Werner, F. (2010). The challenge of adapting marine social-ecological systems to the additional stress of climate change. *Current Opinion in Environmental Sustainability, 2*(5–6), 356–363. https://doi.org/10.1016/j.cosust.2010.10.004

Petzold, J., & Ratter, B. M. (2015). Climate change adaptation under a social capital approach – An analytical framework for small islands. *Ocean & Coastal Management, 112,* 36–43. https://doi.org/10.1016/j.ocecoaman.2015.05.003

Pincha, C. (2008). *Indian Ocean Tsunami through the gender lens: Insights from Tamil Nadu, India.* Mumbai: Oxfam America & NANBAN Trust. https://bit.ly/2EM1Lgk

Piya, L., Maharjan, K. L., & Joshi, N. P. (2012). Perceptions and realities of climate change among the Chepang communities in rural mid-hills of Nepal. *Journal of Contemporary India Studies: Space and Society, Hiroshima University, 2*(5), 35–50. https://ir.lib.hiroshima-u.ac.jp/files/public/3/33600/20141016193954727163/J-ContempIndiaStud-SpaceSoc-HiroshimaUniv_2_35.pdf

Pörtner, H.-O., Karl, D. M., Boyd, P. W., Cheung, W. W. L., Lluch-Cota, S. E., Nojiri, Y., Schmidt, D. N., & Zavialov, P. O. (2014). Ocean systems. In *Climate change 2014: Impacts, adaptation, and vulnerability. Part A: Global and sectoral aspects. Contribution of working group II to the fifth assessment report of the intergovernmental panel on climate change* [Field, C. B., V. R. Barros, D. J. Dokken, K. J. Mach, M.

D. Mastrandrea, T. E. Bilir, M. Chatterjee, K. L. Ebi, Y. O. Estrada, R. C. Genova, B. Girma, E. S. Kissel, A. N. Levy, S. MacCracken, P. R. Mastrandrea, & L. L. White (Eds.)]. Cambridge University Press, Cambridge, United Kingdom and New York, NY, USA, pp. 411–484

Prater, C., Peacock, W. G., Arlikatti, S., & Grover, H. (2006). Social capacity in Nagapattinam, Tamil Nadu after the December 2004 Great Sumatra earthquake and tsunami. *Earthquake Spectra, 22*(3_suppl), 715–729.

Pretty, J., & Ward, H. (2001). Social capital and the environment. *World Development, 29*(2), 209–227. https://doi.org/10.1016/S0305-750X(00)00098-X

Putnam, R. D. (2000). *Bowling alone: The collapse and revival of American community*. New York: Simon and Schuster.

Putnam, R. D. (2017). Hearing on the State of Social Capital in America. Draft of May 16, 2017. https://www.jec.senate.gov/public/_cache/files/222a1636-e668-4893-b082-418a100fd93d/robert-putnam-testimony.pdf

Qudrat-I Elahi, K., & Lutfor Rahman, M. (2006). Micro-credit and micro-finance: Functional and conceptual differences. *Development in Practice, 16*(5), 476–483. https://doi.org/10.1080/09614520600792481

Raabe, K., Sekher, M., & Birner, R. (2009). *The effects of political reservations for women on local governance and rural service provision*. IFPRI Discussion Paper 00878, Development Strategy and Governance Division, International Food Policy Research Institute.

Racioppi, L., & Rajagopalan, S. (Eds.). (2016). *Women and disasters in South Asia survival, security and development*. London: Routledge India.

Rai, N. D., Benjaminsen, T. A., Krishnan, S., & Madegowda, C. (2019). Political ecology of tiger conservation in India: Adverse effects of banning customary practices in a protected area. *Singapore Journal of Tropical Geography, 40*(1), 124–139. https://doi.org/10.1111/sjtg.12259

Rai, P. (2013). *Political representation and empowerment: Women in local government institutions in Bihar, India*. Research report. Stockholm University, Department of Political Science.

Rajasekhar, D. (2004). The interface between gram panchayats and NGOs: Issues, strategies and ways forward, *Journal of Karnataka Studies, 2*, May–October 2004.

Rajshekar, M. (2016). *Why the fishermen of Tamil Nadu can no longer get fish*. Scroll. in. https://scroll.in/article/808960/why-tamil-nadus-fisherfolk-can-no-longer-find-fish

Ram, K. (1988). *Mukkuvar women: The sexual contradictions of capitalist development in a south-Indian fishing community* (Doctoral dissertation). The Australian National University. https://doi.org/10.25911/5d70efdf2abbb

Ramachandran, A., Praveen, D., Radhapriya, P., Divya, S., Remya, K., & Palanivelu, K. (2016). Vulnerability and adaptation assessment a way forward for sustainable sectoral development in the purview of climate variability and change: Insights from the coast of Tamil Nadu, India. *International Journal of Global Warming, 10*(1/2/3), 307. http://dx.doi.org/10.1504/ijgw.2016.077896

Ramachandran, A., Saleem Khan, A., Palanivelu, K., Prasannavenkatesh, R.,& Jayanthi, N. (2017). Projection of climate change–induced sea-level rise for the coasts of Tamil Nadu and Puducherry, India using SimCLIM: A first step towards planning adaptation policies. *Journal of Coastal Conservation*, 21(6), 731–742. https://doi.org/10.1007/s11852-017-0532-6

Raman, V (2005). *Identity formation, nationhood and women: An overview of issues*. Occasional Paper No. 42. New Delhi: Centre for Women's Development Studies.

Ramana Murthy, M., Reddy, N., Pari, Y., Usha, T., & Mishra, P. (2012). Mapping of seawater inundation along Nagapattinam based on field observations. *Natural Hazards*, *60*(1), 161–179. http://dx.doi.org/10.1007/s11069-011-9950-1

Ramu, S., Anandaraj, T., Elaiyaraja, C., & Panneerselvam, A. (2015). Check list of marine fish from Nagapattinam coastal waters, southeast coast of India. *International Journal of Fisheries and Aquatic Studies*, *2*(6), 193–197.

Rao, N., Mishra, A., Prakash, A., Singh, C., Qaisrani, A., Poonacha, P., Vincent, K., & Bedelian, C. (2019). A qualitative comparative analysis of women's agency and adaptive capacity in climate change hotspots in Asia and Africa. *Nature Climate Change*, *9*(12), 964–971. https://doi.org/10.1038/s41558-019-0638-y

Rao, A. D., Upadhaya, P., Ali, H., Pandey, S., & Warrier, V. (2020). Coastal inundation due to tropical cyclones along the east coast of India: An influence of climate change impact. *Natural Hazards*, *101*, 39–57. https://doi.org/10.1007/s11069-020-03861-9

Ray-Bennett, N. S. (2009). The influence of caste, class and gender in surviving multiple disasters: A case study from Orissa, India. *Environmental Hazards*, *8*(1), 5–22. https://doi.org/10.3763/ehaz.2009.0001

Ray-Bennett, N. S. (2010). Multiple disasters and micro-credit: Can it reduce women's vulnerability? the case of Tarasahi in Orissa, India. *Disasters*, *34*(1), 240–260.

Raymond, C. M., Fazey, I., Reed, M. S., Stringer, L. C., Robinson, G. M., & Evely, A. C. (2010). Integrating local and scientific knowledge for environmental management. *Journal of Environmental Management*, *91*(8), 1766–1777. https://doi.org/10.1016/j.jenvman.2010.03.023

Reed, M. S., Podesta, G., Fazey, I., Geeson, N., Hessel, R., Hubacek, K., ... & Thomas, A. D. (2013). Combining analytical frameworks to assess livelihood vulnerability to climate change and analyse adaptation options. *Ecological Economics*, *94*, 66–77. https://doi.org/10.1016/j.ecolecon.2013.07.007

Rege, S., Devika, J., Kannabiran, K., John, M. E., Swaminathan, P., & Sen, S. (2013). Intersections of Gender and Caste, *Economic and Political Weekly*, 48(18), 35–36.

Regnier, P. (2007). *From post-tsunami emergency assistance to livelihood recovery in South India: Exploring the contribution of micro-entrepreneurship initiatives in the Gulf of Mannar, India*. Geneva: International Federation terre des hommes & Terre des Hommes Suisse. In collaboration with Graduate Institute of Development Studies & Graduate Institute of International Studies.

Rehman, S., Sahana, M., Kumar, P., Ahmed, R., & Sajjad, H. (2021). Assessing hazards induced vulnerability in coastal districts of India using site-specific indicators: An integrated approach. *GeoJournal*, *86*(5), 2245–2266. https://doi.org/10.1007/s10708-020-10187-3

Reid, H., Alam, M., Berger, R., Cannon, T., Huq, S., & Milligan, A. (2009). *Community-based adaptation to climate change: An overview* (pp. 11–38). Participatory Learning and Action. London: IIED.

Revathi, R. (2015, July). Tsunami-hit farmers of Nagapattinam against prawn industry. *Down to Earth*. https://www.downtoearth.org.in/news/tsunamihit-farmers-of-nagapattinam-against-prawn-industry-8884

Ribot, J. C. (2011). Vulnerability before adaptation: Toward transformative climate action. *Global Environmental Change*, *21*(4), 1160–1162. https://doi.org/10.1016/j.gloenvcha.2011.07.008

Ribot, J. C., Najam, A., & Watson, G. (1996). Climate variation, vulnerability and sustainable development in the semi-arid tropics. In J. C. Ribot, A. R. Magalhaes, & S. Panagides (Eds.), *Climate variability, climate change and social vulnerability in the semi-arid tropics* (pp. 13–51). Cambridge: Cambridge University Press.

Ribot, J. C. (2010). Vulnerability does not just fall from the sky: Toward multi-scale pro-poor climate policy. In R. Mearns, & A. Norton (Eds.), *Social dimensions of climate change: Equity and vulnerability in a warming world* (pp. 45–74). Washington, DC: The World Bank.

Richards, P., & Roberts, B. (1998, May). Social networks, social capital, popular organizations, and urban poverty: A research note. In *Seminar on Urban Poverty sponsored by ALOP and the World Bank, Rio de Janeiro, May* (pp. 14–16).

Rocheleau, D., Thomas-Slayter, B., & Wangari, E. (1996). *Feminist political ecology: Global issues and local experiences*. London: Routledge. https://doi.org/10.2307/3060380.

Rockenbauch, T., & Sakdapolrak, P. (2017). Social networks and the resilience of rural communities in the Global South: A critical review and conceptual reflections. *Ecology and Society, 22*(1). https://www.jstor.org/stable/26270110

Rodriguez, S., Balasubramanian, G., Shiny, M. P., Duraiswamy, M., & Jaiprakash, P. (2008). *Beyond the Tsunami: Community perceptions of resources, policy, and development, post-tsunami interventions, and community institutions in Tamil Nadu, India*. Bangalore: United Nations Development Program and United Nations Team for Recovery Support, Chennai & Ashoka Trust for Research in Ecology and the Environment.

Rosenthal, M. (2016). Qualitative research methods: Why, when, and how to conduct interviews and focus groups in pharmacy research. *Currents in Pharmacy Teaching & Learning, 8*(4), 509–516. https://doi.org/10.1016/j.cptl.2016.03.021

Rotberg, F. (2010). Social networks and adaptation in rural Bangladesh. *Climate and Development, 2*(1), 65–72. https://doi.org/10.3763/cdev.2010.0031

Rotberg, F. (2012). Social networks, brokers, and climate change adaptation. A Bangladeshi case. *Journal of International Development, 25*(5), 599–608. https://doi.org/10.1002/jid.2857

Roy, D. (2020). 'On the horns of a dilemma'! Climate change, forest conservation and the marginal people in Indian sundarbans. *Forum for Development Studies, 47*(2). https://doi.org/10.1080/08039410.2020.1786452

Sagayaraj, A. (2013). Self-help group and women empowerment: A case study of the disaster survivors of Nagapattinam, Tamil Nadu, South India. *Annual Papers of the Anthropological Institute, Nanzan University, 3*, 152–168.

Salagrama, V. (2006). *Trends in poverty and livelihoods in coastal fishing communities of Orissa State, India*. FAO Fisheries Technical Paper 490. Rome: Food and Agriculture Organization of the United Nations.

Salagrama, V. (2012). *Climate change and fisheries: Perspectives from small-scale fishing communities in India on measures to protect life and livelihood*. International Collective in Support of Fishworkers. https://aquadocs.org/bitstream/handle/1834/27415/Climate_Change_Full.pdf?sequence=1&isAllowed=y

Salagrama, V., & Koriya, T. (2008). *Assessing opportunities for livelihood enhancement and diversification in coastal fishing communities of Southern India*. Chennai: United Nations Team for Tsunami Recovery Support, UN India. http://www.indiaenvironmentportal.org.in/files/Fisheriestudy-June2008-FINAL.pdf

Samas. (2015). *Neer nilam vanam kadal*. Chennai: Kasturi & Sons Limited. (in Tamil).

Samuel, J. (2016). *Tamil Nadu fishermen leverage smartphone technology to boost fish yields*. Village Square. https://www.villagesquare.in/tamil-nadu-fishermen-leverage-smartphone-technology-boost-fish-yields/

Santha, S. (2008). Local ecological knowledge and fisheries management: A study among riverine fishing communities in Kerala, India. *Local Environment, 13*(5), 423–435. https://doi.org/10.1080/13549830701809726.

Sanyal, N. (2006). *Political ecology of environmental crises in Bangladesh* (Masters thesis). University of Durham.

Sapovadia, V. K. (2004, July). Fisherman cooperatives: A tool for socio-economic development. In *International Institute of Fisheries Economics & Trade Conference*. https://ssrn.com/abstract=954986

Saxena, S., Geethalakshmi, V., & Lakshmanan, A. (2013). Development of habitation vulnerability assessment framework for coastal hazards: Cuddalore coast in Tamil Nadu, India – A case study. *Weather and Climate Extremes, 2*, 48–57. https://doi.org/10.1016/j.wace.2013.10.001

Schell, C. (1992). The value of the case study as a research strategy. *Manchester Business School, 2*(1), 1–15.

Schipper, L., & Pelling, M. (2006). Disaster risk, climate change and international development: Scope for, and challenges to, integration. *Disasters, 30*(1), 19–38. https://doi.org/10.1111/j.1467-9523.2006.00304.x

Scoones, I. (1998). Sustainable rural livelihoods: A framework for analysis. IDS working paper, 72. Brighton: IDS.

Scott, R. W. (1995). Institutions and organizations. *Ideas and Interests*. Thousand Oaks: SAGE Publications.

Selltiz, C. Wrightsman, L. C., & Cook, S. W. (1976). *Research methods in social relations* (3rd ed.). New York: Holt Rinehart & Winston.

Selvam, V., Ravishankar, T., Karunagaran, V. M., Ramasubramanian, R., Eganathan, P., & Parida, A. K. (2005). *Toolkit for establishing coastal bioshield*. Chennai: M. S. Swaminathan Research Foundation.

Sen, A. (1999). *Development as freedom*. Oxford: Oxford University Press.

Seneviratne, S. I., Nicholls, N., Easterling, D., Goodess, C. M., Kanae, S., Kossin, J., Luo, Y., Marengo, J., McInnes, K., Rahimi, M., Reichstein, M., Sorteberg, A., Vera, C., & Zhang, X. (2012). Changes in climate extremes and their impacts on the natural physical environment. In *Managing the risks of extreme events and disasters to advance climate change adaptation* [Field, C. B., V. Barros, T. F. Stocker, D. Qin, D. J. Dokken, K. L. Ebi, M. D. Mastrandrea, K. J. Mach, G.-K. Plattner, S. K. Allen, M. Tignor, & P. M. Midgley (Eds.)]. *A Special Report of Working Groups I and II of the Intergovernmental Panel on Climate Change (IPCC)*. Cambridge University Press, Cambridge, UK, and New York, NY, USA, pp. 109–230.

Shaffril, H., Abu Samah, A., & D'Silva, J. (2017). Climate change: Social adaptation strategies for fishermen. *Marine Policy, 81*, 256–261. https://doi.org/10.1016/j.marpol.2017.03.031

Shank, G. (2002). *Qualitative research. A personal skills approach*. New Jersey: Merril Prentice Hall.

Shanthi, B., Mahalakshmi, P., & Chandrasekaran, V. S. (2017). Assessment pre and post tsunami impacts on the livelihoods of coastal women using socio- economic and gender analysis (SEAGA). *Asian Fisheries Science Special Issue, 30S*, 199–217

Sharma, U., & Patt, A. G. (2012). Disaster warning response: The effects of different types of personal experience. *Natural Hazards, 60*(2), 409–423. https://doi.org/10.1007/s11069-011-0023-2

Sharma, U., Patwardhan, A., & Parthasarathy, D. (2009). Assessing adaptive capacity to tropical cyclones in the East coast of India: A pilot study of public response to cyclone warning information. *Climatic Change, 94*(1–2), 189–209. https://doi.org/10.1007/s10584-009-9552-z

Sharma, U., Patwardhan, A., & Patt, A. G. (2013). Education as a determinant of response to cyclone warnings: Evidence from coastal zones in India. *Ecology and Society, 18*(2). https://doi.org/10.5751/es-05439-180218

Shaw, R., Sharma, A., & Takeuchi, Y. (Eds.). (2009). *Indigenous knowledge and disaster risk reduction: From practice to policy*. New York: Nova Science Publishers.
Sheik Mujabar, P., & Chandrasekar, N. (2013). Coastal erosion hazard and vulnerability assessment for southern coastal Tamil Nadu of India by using remote sensing and GIS. *Natural Hazards*, 69(3), 1295–1314. https://doi.org/10.1007/s11069-011-9962-x
Sheth, A., Sanyal, S., Jaiswal, A., & Gandhi, P. (2006). Effects of the December 2004 Indian Ocean Tsunami on the Indian mainland. *Earthquake Spectra*, 22(S3), 435–473. https://doi.org/10.1193/1.2208562
Shrestha, G., Joshi, D., & Clement, F. (2019). Masculinities and hydropower in India: A feminist political ecology perspective. *International Journal of the Commons*, 13(1), 130–152.
Shukla, G., Kumar, A., Pala, N. A., & Chakravarty, S. (2016). Farmers perception and awareness of climate change: A case study from Kanchandzonga Biosphere Reserve. *Environment, Development and Sustainability*, 18(4), 1167–1176. https://doi.org/10.1007/s10668-015-9694-2
Shulman, D. (2016). *Tamil: A biography*. Cambridge, MA: Belknap Press of Harvard University Press
Silverman, D. (2000). *Doing qualitative research: A practical handbook*. Thousand Oaks, CA: Sage.
Singh, C., Deshpande, T., & Basu, R. (2017). How do we assess vulnerability to climate change in India? A systematic review of literature. *Regional Environmental Change*, 17(2), 527–538. https://doi.org/10.1007/s10113-016-1043-y
Singh, R., & Goswami, D. (2010). Evolution of panchayats in India. *Kurukshetra*, 58(12), 3–8.
Small, L. (2007). The sustainable rural livelihoods approach: A critical review. *Canadian Journal of Development Studies / Revue Canadienne D'études Du Développement*, 28(1), 27–38. https://doi.org/10.1080/02255189.2007.9669186
Smit, B., Burton, I., Klein, R. J., & Street, R. (1999). The science of adaptation: A framework for assessment. *Mitigation and Adaptation Strategies for Global Change*, 4(3), 199–213. https://doi.org/10.1023/A:1009652531101
Smit, B., Burton, I., Klein, R. J., & Wandel, J. (2000). An anatomy of adaptation to climate change and variability. In S. M. Kane, & G. W. Yohe (Eds.), *Societal adaptation to climate variability and change* (pp. 223–251). Springer, Dordrecht.
Smit, B., & Pilifosova, O. (2001). Adaptation to climate change in the context of sustainable development and equity. In J. J. McCarthy, O. Canziani, N. A. Leary, D. J. Dokken, & K. S. White (Eds.), *Climate change 2001: Impacts, adaptation, and vulnerability*. Contribution of working group II to the third assessment report of the intergovernmental panel on climate change. Cambridge, UK: Cambridge University Press, pp. 879–906.
Smit, B., & Wandel, J. (2006). Adaptation, adaptive capacity and vulnerability. *Global Environmental Change*, 16(3), 282–292. https://doi.org/10.1016/j.gloenvcha.2006.03.008
Smithers, J., & Smit, B. (1997). Human adaptation to climatic variability and change. *Global Environmental Change*, 7(2), 129–146. https://doi.org/10.1016/S0959-3780(97)00003-4
SoE. (2005). *State of environment report of Tamil Nadu*. Ministry of Environment and Forest, Government of India.
Spencer, R. W. (2008). *Climate confusion: How global warming hysteria leads to bad science, pandering politicians and misguided policies that hurt the poor*. New York: Encounter Books.
Sridhar, A., Arthur, R., Goenka, D., Jairaj, B., Mohan, T., Rodriguez, S., & Shanker, K. (2006). *Review of the Swaminathan committee report on the CRZ notification* (32 p). New Delhi, India: UN Development Programme.

Srinivasan, R., & Nagarajan, K. (2005). Tsunami survey in the Chennai–Nagapattinam segment Tamil Nadu coast. In Dasgupta, S. (Ed.), *Sumatra Andaman earthquake and tsunami 26 December 2004* (pp. 197–216). Kolkata: Geological Survey of India special publication no. 89.

Stake, R. (2010). *Qualitative research*. New York: Guilford Press.

Starman, A. B. (2013). The case study as a type of qualitative research. *Journal of Contemporary Educational Studies/Sodobna Pedagogika, 64*(1), 28–43.

State of Environment Report for Tamil Nadu. (2016). *State of environment report for Tamil Nadu*. Centre of Excellence in Environmental Economics, Madras School of Economics, Chennai. http://www.tnenvis.nic.in/WriteReadData/LatestNewsData/SoERTN_MSE_Final_Jan_27_2016%20(1).pdf

Stern, N. (2007). *The economics of climate change – The Stern review*. Cambridge, UK: Cambridge University Press.

Sternberg, T. (2017). *Climate hazard crises in Asian societies and environments*. Abingdon and New York. Routledge.

Stirrat, R. L. (2004). Yet another 'magic bullet': The case of social capital. *Aquatic Resources, Culture and Development, 1*(1), 25–33. https://doi.org/10.1079/ARC20044

Stoll, J. S., Fuller, E., & Crona, B. I. (2017). Uneven adaptive capacity among fishers in a sea of change. *PLoS One, 12*(6), e0178266. https://doi.org/10.1371/journal.pone.0178266

Strauss, A. (1987). *Qualitative analysis of social science*. Cambridge: Cambridge University Press.

Subramanian, A. (2009). *Shorelines: Space and rights in South India*. Stanford: Stanford University Press.

Subramanian, S. (2011). *Following fish*. Gurgaon: Penguin Books Limited.

Sultana, F (2015) Emotional political ecology. In Bryant, R. (Ed.), *The international handbook of political ecology* (pp. 633–645). London: Edward Elgar. https://doi.org/10.4337/9780857936172.00056.

Sultana, F. (2021). Political ecology 1: From margins to center. *Progress in Human Geography, 45*(1), 156–165. https://doi.org/10.1177%2F0309132520936751

Sundar, A. (2010). *Capitalist transformation and the evolution of civil society in a South Indian fishery* (Doctoral dissertation). University of Toronto.

Sundar, V., & Sundaravadivelu, R. (2005). Protection measures for Tamil Nadu Coast. *Public Works Department, Government of Tamil Nadu*. https://tnsdma.tn.gov.in/app/webroot/img/document/library/15-Protection-Measures-For-Tamil-nadu-cost.pdf

Swamy, Raja. (2009). The fishing community and heritage tourism in Tarangambadi. In E. Fihl, & A. R. Venkatachalapathy (Eds.), *Indo-Danish cultural encounters in Tranquebar: Past and present*. Special issue of *Review of Development and Change*, 14 (1–2), 197–226. http://natmus.dk/fileadmin/user_upload/natmus/forskning/dokumenter/Tranquebar/RDC_XIV_Tranquebar.pdf

Swamy, R. H. (2011). Disaster capitalism: Tsunami reconstruction and neoliberalism in Nagapattinam, South India (Doctoral dissertation). University of Texas at Austin. http://hdl.handle.net/2152/ETD-UT-2011-05-3461

Swamy, R. H. (2014). Risk and opportunity in post-tsunami Tharangamapadi. In E. Fihl, & A. R. Venkatachalapathy (Eds.), *Beyond tranquebar: Grappling across cultural borders in South India*. Orient Blackswan.

Swamy, R. (2018). Fishers, Vulnerability, and the Political Economy of Dispossession and Reconstruction in Post-Tsunami Tamil Nadu. *Individual and Social Adaptations to Human Vulnerability* (*Research in Economic Anthropology*), *38*, 103–126. Bingley: Emerald Publishing Limited. https://doi.org/10.1108/S0190-128120180000038006

Swamy, R. H. (2021). *Building back better in India: Development, NGOs, and artisanal fishers after the 2004 tsunami*. Tuscaloosa: University of Alabama Press.

Swathi Lekshmi, P. S., Dineshbabu, A. P., Purushottama, G. B., Thomas, S., Sasikumar, G., Rohit, P., Vivekanandan, E., & Zacharia, P. U. (2013). *Indigenous Technical Knowledge (ITKs') of Indian Marine fishermen with reference to climate change*. Kochi: Central Marine Fisheries Research Institute. http://eprints.cmfri.org.in/9455/1/Indigenous_Technical_Knowledge_%28ITKs%E2%80%99%29_of.pdf

Tamil Nadu Fishermen Welfare Board. (n.d.). *Tamil Nadu fishermen welfare board*. Government of Tamil Nadu. https://tnfwb.tn.gov.in/index.php/User/login

Taylor, M. (2014). *The political ecology of climate change adaptation: Livelihoods, agrarian change and the conflicts of development*. London: Routledge.

Tearfund. (2008). *Linking climate change adaptation and disaster risk reduction*. London: Venton & La Trobe.

Termeer, C., Biesbroek, R., & van den Brink, M. (2012). Institutions for adaptation to climate change: Comparing national adaptation strategies in Europe. *European Political Science*, *11*(1), 41–53. https://doi.org/10.1057/eps.2011.7

Thamizoli, P., & Prabhakar, I (2009). Traditional governance system of fishing communities in Tamil Nadu, India: Internal mandate, interfacing and integrating development. In R. Shaw., & R. R. Krishnamurthy (Eds.), *Communities in coastal zone management* (p. 16). Singapore: Research Publishing Services.

The Hunger Project. (n.d.). *The reports of Hunger Projects*. https://thp.org/

The New Indian Express. (2021a, September 8). *After eight years, Tamil Nadu set to finalise climate change action plan*. The New Indian Express. https://www.newindianexpress.com/states/tamil-nadu/2021/sep/08/after-eight-years-tamil-nadu-set-to-finalise-climate-change-action-plan-2356101.html

The New Indian Express. (2021b, July 19). *Delta districts fisherfolk divided over TN Fishing Regulation Act*. The New Indian Express. https://www.newindianexpress.com/states/tamil-nadu/2021/jul/19/delta-districts-fisherfolk-divided-over-fishing-regulation-act-2332147.html

Thomalla, F., Downing, T., Spanger-Siegfried, E., Han, G., & Rockström, J. (2006). Reducing hazard vulnerability: Towards a common approach between disaster risk reduction and climate adaptation. *Disasters*, *30*(1), 39–48. https://doi.org/10.1111/j.1467-9523.2006.00305.x

Thomas, A. (2004). *Formal and informal Institutions: Gender and participation in the Panchayati Raj* (Doctoral dissertation). Western Michigan University. https://scholarworks.wmich.edu/s/1145

Thomas, B. K., Muradian, R., de Groot, G., & de Ruijter, A. (2010). Resilient and resourceful? A case study on how the poor cope in Kerala, India. *Journal of Asian and African Studies*, *45*(1), 29–45. https://doi.org/10.1177%2F0021909610353580

Thompson, P. (1985). Women in the fishing: The roots of power between the sexes. *Comparative Studies in Society and History*, *27*(1), 3–32. https://doi.org/10.1017/S0010417500013645

Thornton, P. K., Ericksen, P. J., Herrero, M., & Challinor, A. J. (2014). Climate variability and vulnerability to climate change: A review. *Global Change Biology*, *20*(11), 3313–3328. https://doi.org/10.1111/gcb.12581

Thurston, E. & Rangachari, K. (1909). *The castes and tribes of Southern India*. Madras: Government Press.

Tibby, J., Lane, M. B., & Gell, P. A. (2007). Local knowledge and environmental management: A cau- tionary tale from Lake Ainsworth, New South Wales, Australia. *Environmental Conservation, 34*(04), 334–341. https://doi.org/10.1017/s037689290700433x

TNAU Agritech Portal. (2015). *Fishing crafts.* Tamil Nadu Agricultural University Portal. https://agritech.tnau.ac.in/fishery/fish_pht_tech_grafts.html

TNSDMA. (2016). *Tamil Nadu state disaster management plan 2016.* Tamil Nadu State Disaster Management Authority (TNSDMA). Government of Tamil Nadu. https://tnsdma.tn.gov.in/app/webroot/img/document/tnsdma-2016.pdf

TNSDMA. (2018). *State disaster management perspective plan 2018–2030.* Tamil Nadu State Disaster Management Authority (TNSDMA). Government of Tamil Nadu. https://tnsdma.tn.gov.in/app/webroot/img/document/SDMP-29-08.pdf

Tol, R. S., Fankhauser, S., & Smith, J. B. (1998). The scope for adaptation to climate change: What can we learn from the impact literature? *Global Environmental Change, 8*(2), 109–123. https://doi.org/10.1016/S0959-3780(98)00004-1

Tompkins, E., & Adger, W. (2004). Does adaptive management of natural resources enhance resilience to climate change? *Ecology and Society, 9*(2). http://www.ecologyandsociety.org/vol9/iss2/art10/

Truelove, Y. (2011). (Re-) Conceptualizing water inequality in Delhi, India through a feminist political ecology framework. *Geoforum, 42*(2), 143–152. https://doi.org/10.1016/j.geoforum.2011.01.004

UNEP. (2009). *South Asia environment outlook 2009.* UNEP, SAARC and DA. https://wedocs.unep.org/20.500.11822/8004

UNFCCC. (2006). *Climate change: Impacts, vulnerabilities, and adaptation in developing countries.* United Nations Framework Convention on Climate Change. https://unfccc.int/resource/docs/publications/impacts.pdf

van Oldenborgh, G. J., Philip, S., Kew, S., van Weele, M., Uhe, P., Otto, F., Singh, R., Pai, I., Cullen, H., & AchutaRao, K. (2018). Extreme heat in India and anthropogenic climate change. *Natural Hazards and Earth System Sciences, 18*(1), 365–381. https://doi.org/10.5194/nhess-18-365-2018

Vedwan, N. (2006). Culture, climate and the environment: Local knowledge and perception of climate change among apple growers in northwestern India. *Journal of Ecological Anthropology, 10*, 4–15.

Venkatasubramanian, K., & Ramnarain, S. (2018). Gender and adaptation to climate change: Perspectives from a pastoral community in Gujarat, India. *Development and Change, 49*(6), 1580–1604. https://doi.org/10.1111/dech.12448

Vinke, K., Martin, M. A., Adams, S., Baarsch, F., Bondeau, A., Coumou, D., Donner, R. V., Menon, A., Perrette, M., Rehfeld, K., & Robinson, A., 2017. Climatic risks and impacts in South Asia: Extremes of water scarcity and excess. *Regional Environmental Change, 17*(6), 1569–1583.

Viswanathan, P. K. (2013). Conservation, restoration, and management of mangrove wetlands against risks of climate change and vulnerability of coastal livelihoods in Gujarat. In S. Nautiyal, K. S. Rao, H. Kaechele, K. V. Raju, & R. Schaldach (Eds.), *Knowledge systems of societies for adaptation and mitigation of impacts of climate change* (pp. 423–441). Berlin: Springer-Verlag.

Vivekanandan, E. (2010). Impact of climate change in the indian marine fisheries and the potential adaptation options. In *Coastal fishery resources of India–Conservation and sustainable utilisation* (pp. 169–185). Society of Fisheries Technologists.

Vivekanandan, E. (2011). *Marine fisheries policy brief 3; climate change and Indian marine fisheries* (Vol. 105, pp. 1–97). CMFRI Special Publication. http://eprints.cmfri.org.in/8440/1/CMFRI_SP_105.pdf

Vivekanandan, E., & Jeyabaskaran, R. (2010). Impact and adaptation options for Indian marine fisheries to climate change. In G. S. L. H. V. Prasad Rao (Ed.), *Climate change adaptation strategies in agriculture and allied sectors* (pp. 107–117). New Delhi: Scientific Publishers.

Vivekananda, J., Schilling, J., Mitra, S., & Pandey, N. (2014). On shrimp, salt and security: Livelihood risks and responses in South Bangladesh and East India. *Environment, Development and Sustainability*, *16*(6), 1141–1161. https://doi.org/10.1007/s10668-014-9517-x

Voss, M. (2008). The vulnerable can't speak. An integrative vulnerability approach to disaster and climate change research. *Behemoth-A Journal on Civilisation*, *1*(3), 39–56. https://doi.org/10.6094/behemoth.2008.1.3.730

Walker, R. (ed.). (1985). *Applied qualitative research*. Aldershot: Gower.

Walther, G. R., Post, E., Convey, P., Menzel, A., Parmesan, C., Beebee, T. J., Fromentin, J. M., Hoegh-Guldberg, O., & Bairlein, F. (2002). Ecological responses to recent climate change. *Nature*, *416*(6879), 389–395. https://doi.org/10.1038/416389a

Wang, J., Wang, Y., Li, S., & Qin, D. (2016). Climate adaptation, institutional change, and sustainable livelihoods of herder communities in northern Tibet. *Ecology and Society*, *21*(1). https://doi.org/10.5751/es-08170-210105

Warner, K. (2010). Global Environmental change and migration: Governance challenges. *Global Environmental Change*, *20*(3), 402–413. https://doi.org/10.1016/j.gloenvcha.2009.12.001

Warner, K., Ehrhart, C., Sherbinin, A. D., Adamo, S., & Chai-Onn, T. (2009). In search of shelter: Mapping the effects of climate change on human migration and displacement. *A policy paper prepared for the 2009 Climate Negotiations*. Bonn, Germany. United Nations University, CARE, and CIESIN-Columbia University and in close collaboration with the European Commission "Environmental Change and Forced Migration Scenarios Project", the UNHCR, and the World Bank

Weeratunge, N., Béné, C., Siriwardane, R., Charles, A., Johnson, D., Allison, E. H., Nayak, P. K., & Badjeck, M. C. (2014). Small-scale fisheries through the wellbeing lens. *Fish and Fisheries*, *15*(2), 255–279. https://doi.org/10.1111/faf.12016

Wheaton, E. E., & Maciver, D. C. (1999). A framework and key questions for adapting to climate variability and change. *Mitigation and Adaptation Strategies for Global Change*, *4*(3), 215–225. https://doi.org/10.1023/A:1009660700150

Wheeler, D. (2011). *Quantifying vulnerability to climate change: Implications for adaptation assistance*. Center for Global Development. https://www.cgdev.org/sites/default/files/1424759_file_Wheeler_Quantifying_Vulnerability_FINAL.pdf

Widdowson, F., & Howard, A. (2008). *Disrobing the aboriginal industry: The deception behind indigenous cultural preservation*. Montreal: McGill-Queen's University Press.

Wikipedia Commons. (2015). *TN taluks Nagapattinam*. Author: BishkekRocks. https://commons.wikimedia.org/wiki/File:TN_Taluks_Nagapattinam.png

Wikipedia Tamil. (n.d.). நக்கீரர், சங்கப்புலவர். https://ta.wikipedia.org/wiki/நக்கீரர்,_சங்கப்புலவர்

Williams, M. J. (2016). How are fisheries and aquaculture institutions considering gender issues. *Asian Fisheries Sciences S*, *29*, 21–48. https://www.asianfisheriesociety.org/publication/downloadfile.php?id=1109&file=Y0dSbUx6QXhNekl3TXpjd01ERTBOemczTXprMk1qZ3VjR1Jt

Williams, T., & Hardison, P. (2013). Culture, law, risk and governance: Contexts of traditional knowledge in climate change adaptation. *Climate Change*, *120*, 531–544. https://doi.org/10.1007/s10584-013-0850-0.

Wisner, B. (2003). Disaster risk reduction in megacities: Making the most of human and social capital. In A. Kreimer, M. Arnold, & A. Carlin (Eds.), *Building Safer Cities: The Future of Disaster Risk* (pp. 181–96). Washington, DC: The World Bank.

Wisner, B. (2009). Local knowledge and disaster risk reduction. Keynote. *Side meeting on indigenous knowledge, Global platform for disaster reduction, Geneva*. University College London, England.

WMO. (2019). *Gendered Impacts of Weather and Climate: Evidence from Asia, Pacific and Africa*. Capstone Project Research Report. World Meteorological Organization. https://library.wmo.int/doc_num.php?explnum_id=10106

Wolf, J. (2011). Climate change adaptation as a social process. In J. D. Ford, & L. Berrang-Ford (Eds.), *Climate change adaptation in developed nations: From theory to practice* (pp. 21–32). Dordrecht: Springer.

Wong, P. P., Losada, I. J., Gattuso, J.-P., Hinkel, J., Khattabi, A., McInnes, K. L., Saito, Y., & Sallenger, A. (2014). Coastal systems and low-lying areas. In C. B. Field, V. R. Barros, D. J. Dokken, K. J. Mach, M. D. Mastrandrea, T. E. Bilir, M. Chatterjee, K. L. Ebi, Y. O. Estrada, R. C. Genova, B. Girma, E. S. Kissel, A. N. Levy, S. MacCracken, P. R. Mastrandrea, & L. L. White (Eds.), *Climate change 2014: Impacts, adaptation, and vulnerability. Part A: global and sectoral aspects. Contribution of Working Group II to the Fifth assessment report of the intergovernmental panel on climate change* (pp. 361–409). Cambridge, United Kingdom and New York, NY: Cambridge University Press.

Woolcock, M., & Narayan, D. (2000). Social capital: Implications for development theory, research, and policy. *The World Bank Research Observer*, *15*(2), 225–249. https://doi.org/10.1093/wbro/15.2.225

World Bank. (1991). *Small-scale fisheries: Research needs*, World Bank Technical Paper Number 152, Fisheries Series, Washington, DC: World Bank Publications.

World Bank. (2007). *Vulnerability Reduction of Coastal Communities (VRCC), Environmental and Social Management Framework*. Government of Tamil Nadu. https://documents1.worldbank.org/curated/en/482061468035979271/pdf/E1126v5ETRP1ESMF0revised2VRCC2.pdf

World Bank. (2010). *Indian Marine Fisheries: Issues, Opportunities and Transitions for Sustainable Development Report*. (Report No: 54259-IN). http://documents1.worldbank.org/curated/en/513221468040751464/pdf/542590ESW0whit0ries0Report00PUBLIC0.pdf

World Bank. (2013). *Turn Down the Heat: Climate Extremes, Regional Impacts, and the Case for Resilience*. A report for the World Bank by the Potsdam Institute for Climate Impact Research and Climate Analytics. World Bank. https://www.worldbank.org/content/dam/Worldbank/document/Full_Report_Vol_2_Turn_Down_The_Heat_%20Climate_Extremes_Regional_Impacts_Case_for_Resilience_Print%20version_FINAL.pdf

Yin, R. (2003). *Case study research: Design and methods* (3rd ed.). Thousand Oaks, CA: Sage.

Yin, R. K. (2011). *Qualitative research from start to finish*. New York: The Guilford Press.

Yohe, G., & Tol, R. S. (2002). Indicators for social and economic coping capacity – Moving toward a working definition of adaptive capacity. *Global Environmental Change*, *12*(1), 25–40. https://doi.org/10.1016/S0959-3780(01)00026-7

Zacharia, P. U., Dineshbabu, A. P.,Thomas, S., Kizhakudan, S. J.,Vivekanandan, E. et al. (2016). *Relative vulnerability assessment of Indian marine fishes to climate change using impact and adaptation attributes.* CMFRI Special Publication (125), (CMFRI-NICRA Publication No. 5), Kochi: CMFRI.

Zvelebil, K. V. (1974). *A history of Indian literature* (Vol 10), Jan Gonda (Ed.). Wiesbaden: Otto Harrassowitz.

Index

Page numbers in *italics* denote illustrations, **bold** a table and with n, an endnote.

2004 Indian Ocean Tsunami disaster: coastal erosion, state adaptions 128–129; free-tsunami dwellings, inadequacies 88; Nagapattinam district, relief schemes 48–49, 50, *51*, 52, *53*; Nagapattinam district's vulnerability 12, 39, 40; Tamil Nadu relief schemes 39, 41, 122; water salinity increases 92–93

Abeygunawardena, P. 21
Adaption, Institutions and Livelihoods (AIL) framework 33
adaptive capacity: defining factors 24; educational levels 103–104, 165–166; external actor interventions 165–166, 169; fisherwomen's agency 149–150, 153, 155, 159–160, 172–173; livelihood diversification 103, 126, 167; local institutions and capital factors 32, 98–99, 107–108, 127, 134–138, 169–172; power/social relations 23–24, 105; social vulnerability 87; surukku valai fishers 97–98
adaptive climate planning 175–176
Adger, W.N. 20, 24, 31
Agrawal, A. 31, *31*, 33, 65
Agrawal, Bina 152, 153
Akananuru (poetry anthology) 6–7
Aldrich, D.P. 171
Alexander, C. 85
Allison, E. 32
Alsop, R. 141
Amrith, S.S. 5
Amundsen, H. 170
Ananth Pur, K. 146
Asian Development Bank 39

Assets/Capitals based Adaption, Institutions and Livelihoods (AIL) framework 33, *34*, 35, 162–163
Ayres, J. 23

Babu, S. 41
Balakrishnan, B. 6, 8
Bal, P.K. **38**
Bangladesh 25–26
Bavinck, M.: uur panchayats, governance systems 10–11, 112–113, 114–115, 116, 117; uur panchayats, women's exclusion 147, 160
Bebbington, A. 25
Behera, B. 5
Berardi, A. 65–66
Berkes, F. 65
Berman, R. 33
Bhathal, B. 37
bio-shields 128–129, *128*
Birkmann, J. 22
Bjurström, A. 3
Bogardi, J. 102
Brenkert, A.L. 36
Brulle, R.J. 2
Brundtland Commission 27–29
Bryant, R.L. 26
Byg, A. 65

capitals/assets: Assets/Capitals based Adaption, Institutions and Livelihoods (AIL) framework *34*, 35; capitals/assets 162; marine fishers' studies 33, **33**; social 25; sustainable livelihoods approach 28, 32–33; types 25
case study, advantages 41

Central Marine Fisheries Research Institute 104
Chakraborty, A. 38
Chambers, Robert 27, 28
Chandrasekar, N. 38
Chavadikuppam (study village): climate extremes risks 82, 84, 87–88, 104–105; coastal erosion impacts 99, 133; foreign job migration 102–103; housing and sanitation issues 89; muted politics 118–119; post-2004 Tsunami relief 53; poverty and gender inequalities 91–92; socioeconomic situation 53–54, *54*; state officials, apathetic support 106–107; water salinity, reduction challenges 92–93; women's credit networks 152, 155; women's self-help group 152–153, 155, 156, 158
Cheethalai Chatanar 6, 7
Chinnakottaimedu (study village): climate extremes risks 84–85, 87–88, 104–105; fishing ground changes 79; housing and sanitation issues 89; post-2004 Tsunami relief *51*; poverty and gender inequalities 90–92; socioeconomic situation 50, *51*; state officials, apathetic support 107; women's self-help group 152–153
Chinnamedu (study village): adaptive capacity, women's role 149–150, 153, 155; climate events compensation 94–95; climate extremes risks 87–88, 104–105; coastal erosion impacts 77, 99, 130–131; family remittance support 95–96; fish stock depletion impacts 75, 76; foreign job migration 102–103; housing and sanitation issues 89; muted politics 118–119; poverty and gender inequalities 90–92; socioeconomic situation 53, *54*; state officials, apathetic support 106–107; uur panchayats, capital issues 124; water salinity, reduction challenges 92–93; weather, early warning systems 83; women's self-help group 108, 152–153, 158
Christensen, A.S. 36
Cinner, J. 169
Clark, G.E. 90
climate change: adaption approaches 3–4, 22–23; effects and threats 1, 18; India's coastal vulnerabilities 4–6, 36–37; IPCC key definitions 17, 18–19; marine related challenges 1–2;

scholarly disputes 1, 72; social science research, key insights 2–4; vulnerability and inequality 3, 19; vulnerability concepts 19–20
climate change adaption, concepts and strategies: adaption cycle 23, *23*; approach types 22–23; barriers and limitations 105–107; concept critiques 24; disaster risk reduction 22; foreign job migration 101–103, 126, 166, 167–168; IPCC definition 22; local institutions' involvement 30–32, 31, 99; political ecology discourse 26–27; research gaps, regional and local 174–175; social capital and networks 25; societal/political systems 21, 24, 26–27; stakeholder engagement 21; sustainable livelihoods approach 27–29, 32–33
climate extremes, definition 19
climate variability 17, 18, 61
coastal erosion: adaption discussions 128–133, *128*, *130–132*; adaptive capacity, local institutions' limitations 98–99, 99, 134–135; bio-shields 128–129; dune reconstruction 132–133, 132; impacts 77, 77, 93; post-2004 Tsunami adaptions 128–129; seawall constructions 129–131, 134
Colwell, J.M.N. 143
community-based adaption (CBA) 4, 22–23
Conway, Gordon 27
Corbin, J. 42
Coulthard, S. 125
Crane, T.A. 32
Cutter, S. 19

Devendra Babu, M. 101
Diaz-Cayeros, A. 157
disaster risk reduction: definition 22; external agency facilitation 153, 170, 175–176; fisherwomen's capacity inequalities 108, 120, 144, 149–151, 172–174; indigenous/local knowledge systems 70–71, 84–85; local institution barriers 118, 120–121, 123–124, 149–151; social vulnerability factors 88, 104–105, 108, 136, 167
Divakarannair, N. 33
Dunlap, R.E. 2
Duyne Barenstein, J. 89

Ekstrom, J.A. 105
Emergency Tsunami Reconstruction Project (ETRP) 39, 128
empowerment 141–142
Enarson, E. 174
Eriksen, S.H. 26

Faist, T. 126
Farrington, J. 28, 29
Field, E. 158
Fihl, Esther 11, 41, 114
fisheries and aquaculture 2, 45–47
Fisheries Management for Sustainable Livelihoods (FIMSUL) 112
fishermen cooperative societies: coastal erosion response 132; community responsibilities 102, 119–120, 169; state welfare coordination 100, 136, 170; weather warning alerts 105–106; women's protest support 155
fishermen councils *see* uur panchayats
fisherwomen, challenges and agencies: cooperative societies undermined 100–101, 120, 160; cultural barriers 105, 107–108, 113, 121, 158; disaster risk reduction, capacity inequalities 108, 120, 144, 149–151, 172–174; financial and societal difficulties 90–92, 148–149, 153–154, 158–159; gendered division of labour 142–143, **144**; microfinance, pros and cons 151, 152, 153, 156, 159; NGOs, acceptable support 149–150, 153, 155, 156, 157–158, 159; political capital obstacles 144, 145; research gaps 140–141; sanitation issues 89; self-help groups 108, 121–122, 151–158, 173; uur panchayats, customary exclusion 147–148, 149, 154, 155–156, 160, 169–170
fisherwomen cooperative societies 100–101, 120, 160
Fisherwomen Extension Service 120
fishing castes, Tamil Nadu 8–11, 42–43
Food and Agriculture Organization (FAO) 2, 45–46
Fordham, M. 174
Ford, J. 21, 41
Forsyth, T. 23
Füssel, H.M. 19

Gaillard, J.C. 175
Gandhi, K. 144

gender systems, fishing communities: division of labour 142–143, **144**; external agency influences 157–158, 159–160; gram panchayats undermined 145; local institutions, male-centric 147–148, 149–151; post-2004 Tsunami impacts 144; women's self-help groups 151–157, 159; women's societal obstacles 148–149, 158–159
Glantz, Michael H. 166
global warming, oceans 1–2
Gomathy, N.B. 10, 116
Grafton, R. 171
gram panchayats (local government) 117–119, *118*
grounded theory 41
Guhathakurta, P. 36–37
Gupta, J. 111
Gupta, M. 129

Hapke, H.M. 142
Helmke, G. 30
Holling, C. 20
Hoon, P. 28
Horemans, B. 32
Hovelsrud, G.K. 134
Hulme, M. 1
Hyden, G. 28

Indian coasts, climate change vulnerabilities 4–6, 36–37
Indian National Centre for Ocean Information Services (INCOIS) 165
indigenous/local knowledge systems: adaptive capacity support 165–166; climate change research, value debate 65–66, 163–164; disaster risk reduction capacity 70–71, 84–85; erratic weather risks 81–82, 164–165; fish catches, prediction experience 69–71, **70**; fish declines, fisher perspectives **71**; fishermen's community lingo 66–69; fish stock/catch reduction and climate change 73–79, **80–81**, 81; fish stock reduction and overfishing 76–77; Pattinavar fishermen, key informants 66–68, **67–68**; waning relevance due to erratic weather 83–85, 106, 167; weather, early warning systems 82–83, 84; young fishers disconnection 70–71, 84–85, 104, 165, 168
Infochange 5

Inglin, S. 47
institutions, definitions 29–30, 111
institutions, local *see* local institutions
Intergovernmental Panel on Climate Change (IPCC) 3, 17, 18–19, 20, 22, 24
International Fund for Agricultural Development (IFAD) 39, 122, 152
Islam, M.M. 60, 168

Jacob, V.A. 176
Janakarajan, S. 103
Janssen, M.A. 21
Jaswal, A.K. 37
Jentoft, S. 30
job migration, foreign 62n13, 95–96, 101–103, 126, 166, 167–168
Johnson, D.S. 46
Jones, L. 105
Joshi, P. **38**
Juran, L. 142–143

Kalikoski, D. 97
Kandlikar, M. 166
Karlsson, M. 134
Keck, M. 20–21
Kelly, P.M. 20
Kelman, Ilan 167, 175, 176
Khan, A.S. **38**
King, D. 21
Klein, R.J.T. 19
Kodiyampalayam (study village): climate extremes risks 88; coastal erosion response 133; debt traps 95; disaster management lacking 124–125; economic capital advantages 90, 93; fishing ban impact 78; fish stock depletion impacts 75–76; local institutions, women's deference 149, 150, 154; local knowledge reliance 83; post-2004 Tsunami relief 48–49, 124–125; socioeconomic features 47–48
Koozhayar (study village): coastal erosion response 99, 129; fish stock depletion impacts 75; foreign job migration 102; local knowledge reliance 83; map *49*; post-2004 Tsunami relief 49; socioeconomic situation 49–50, 63n15; surukku valai fishers 96; women's self-help group 108
Koriya, T. 156
Kottaimedu (study village): climate extremes risks 104–105; coastal erosion impacts 77; economic capital advantages 93, 107, 124, 125; map *52*; post-2004 Tsunami relief 50, 52; socioeconomic situation 50, 52; surukku valai fishers 96–98
Kruks-Wisner, G. 153
Kurien, J. 46, 84

Lavell, A.M. 18, 22
Levitsky, S. 30
Li, L. 72
local institutions: adaptive capacity, potential improvement 168–169; Assets/Capitals based AIL framework *34*, 35; case study, key informants 57; climate change adaption, relevance 30–32, *31*, 43–44, 124–126, 136–138; coastal erosion, adaption discussions 128–133, *128*, *130–132*; coastal erosion, adaptive capacity limitations 98–99, *99*, 134–135; evaluation factors 110–111; fishermen cooperative societies 119–120; fisherwomen cooperative societies 120; gram panchayats, functions and influence 117–119, *118*; marine fisheries in India, research 111–112; migrant fishermen, non-regulation 126–127; non-governmental organizations (NGOs) 122–123; political capital bias 118–119, 121; political parties 120–121, 170; political/social capital, capacity barriers 127, 135–136, 137–138, 168, 169–170, 171–172; uur panchayat, functions and governance systems 112–117, *115*; women's representation undermined 117, 147–148, 149–151, 155–156, 160, 170; women's self-help groups 121–122, 124 *see also* uur panchayats (fishermens councils)
Lorber, J. 142
Lorenz, D.F. 21

MacDonald, R. 142–143
Maciver, D.C. 23
Madavamedu (study village): coastal erosion response 99, 130, 131–132, *132*; economic capital advantages 89–90, 93, 107, 124, 125; fishing ban impact 78; fish stock depletion impacts 75; local institutions, women's deference 150, 154; political engagement 119, 127; post-2004

Tsunami relief 52, 53; socioeconomic situation 52; surukku valai fishers 97; women's self-help group dismissed 156–157
Madhanagopal, Devendraraj 77, 176
Mafongoya, P. 31
Mahatma Gandhi National Rural Employment Guarantee Act (MGNREGA) 101, 132
Mahon, R. 41
Malone, E.L. 36
Mandelbaum, D.G. 145–146
marine fishing communities, India: climate change impacts 5–6, 8, 37; mechanized fishing issues 37; study scope and aims 12–13, 163; sustainable livelihoods research 32–33; Tamil Nadu's caste based governance 8–10, *9, 10*
Mazzocchi, F. 65
McGregor, A. 26
McGuire, B. 72
Mexico 156–157
microfinance, women's schemes 151, 152, 153, 154, 156, 159
Miller, F. 26
Mishra, N.K. 142
Mistry, J. 65–66
monsoon patterns, India 5
Moser, S.C. 105
M.S. Swaminathan Research Foundation (MSSRF) 83, 101, 128, 129
Mubaya, C.P. 31
Mukkuvar communities 9, 10–11

Nagapattinam district fishing villages study: 2004 Tsunami disaster 40, *46*; 2004 Tsunami disaster, relief schemes 39, 41; 2004 Tsunami, fishermens' perspective 72; climate change, age linked perceptions 71, 73–74, *74*; climate extremes, risks and events 39, **40**; climatic zone 38; fieldwork challenges 60–61; fishing ban, added difficulties 78–79; fishing resources 38–39; gender representation issue 57, 60–61; government and NGO roles 43–44, 44–45; interviews/group discussions 58–60, **58**; key informants 56–57, **59**, 66–67; shrimp farming and salinity issues 92–93; site selection and categorization 45–47; social vulnerability and climate change 87–92; stakeholder engagement (pilot survey) 42–44; study scope and aims 12–13, 162–163; villages, socioeconomic situation 47–55, 48–49, 51–54
Nakashima, D. 67
Nakkīraṉār 6, 7
Narayanan, K. 36
Narayan, D. 25
Nath, P.K. 5
National Bank for Agriculture and Rural Development (NABARD) 151
National Commission of Women 151
Nayak, Nalini 84, 125, 172
non-governmental organizations (NGOs): fisherwomen cooperative societies 160; fishing communities' involvement 122–123, 135, 138; seawall protests, women's actions 149–150, 153, 155; uur panchayats, cooperation 149–150, 155, 159, 169; uur panchayats, local conflicts 154, 169; women's credit networks 152–153; women's self-help groups 151–154, 156, 157–158, 159
North, D.C. 29

Olsson, L. 61, 67

Paerregaard, K. 79, 81
Pahl-Wostl, C. 30
Palanithurai, G. 117
Panda, S.M. 151
Panditharatne, C. 168
Paravar communities 9
Parthasarathy, D. 104
Parthasarathy, R. 129
Patnaik, U. 36
Patt, A.G. 66
Pattinavar communities 9, 10–11
Pauly, D. 2
Pearse, R. 140
Pelling, M. 23
Perrin, N. 31, 33
Peru 79, 81
Petzold, J. 31
Pilifosova, O. 24
Pincha, C. 142–143
political ecology 26–27
Polk, M. 3
Post Tsunami Sustainable Livelihood Programme 39
Putnam, Robert 25

qualitative research 55–56

Rajshekar, M. 76
Ramana Murthy, M. 40
Ram, K. 143, 149, 154
Rao, A.D. 142
Ratter, B.M. 31
Ray-Bennett, N.S. 153, 159
Reid, H. 23
Reliance Foundation 83, 101
research methodology: case study 41; data collection, key informants 56–57, 59–60, **59**; government and NGO sources 43–44, *44–45*; interviews/group discussions 58–60, **58**; pilot survey of stakeholders 42–43; qualitative research 55–56; sampling strategy 57–58; site selection and categorization 45–47; theological influences 41–42
Reserve Bank of India (RBI) 151
resilience concept 20–21
Ribot, J.C. 20, 24, 87
Rotberg, F. 25
Roue, M. 67

Sagar, A. 166
Sakdapolrak, P. 20–21
Salagrama, V. 5, 112, 140, 156
Salick, J. 65
Sangam literature 6–8
Santha, S. 66
Schröter, D. 66
Scoones, I. 28
Scott, R.W. 29
seawalls 129–131, 134, 171
Sen, A. 141–142
Seneviratne, S.I. 18
Shank, G. 55
Sheik Mujabar, P. **38**
shelter beds 128–129, *128*
Silverman, D. 56
Singh, C. 66
small-scale fisheries, categorization 45–47
Smit, B. 23, 24
social capital, concept 25
social resilience 20–21
social vulnerability and climate change: climate events compensation 93–95, 106–107, 171; coastal erosion impacts 98–99, *99*, 128–133; communication/technology restraints 83–84, 105–106, 165; coping and adaption strategies 100–101; debt traps 95–96; disaster risk reduction capacity, poor 88, 104–105, 108, 123–124, 136, 167; economic capital advantages 89–90, 93, 107, 170–171; educational level issues 103–104; evaluation factors 87, 167; fisherwomen's challenges 89, 90–92, 107–108, 154; free-tsunami dwellings, inadequacies 88; geographic isolation 87–88; local institution challenges 99, 106–107, 118–119, 123–124, 127, 168–172; poverty and gender inequalities 90–92, 107–108, 167; sanitation and wastewater management 89–90; state officials, apathetic support 106–107, 168; surukku valai fishers 96–98, *97*, 167; water salinity issues 92–93
social vulnerability, concept 20
Spencer, R.W. 72
Sridhar, A. 128
Stake, R. 55
Starman, A.B. 55
Stern review 1, 3
Stoll, J.S. 167
Strauss, A. 42
Sundar, Aparna 142, 153
surukku valai (purse net) fishing: adoption, Tamil Nadu 47; climate vulnerability 96–98, *97*, 167; fish stock depletion impacts 75; local disputes and 2020 ban 62n8; nets *98*; overfishing criticisms 76–77; post-2004 Tsunami expansion 52, 63n19
sustainable livelihoods approach: concept and appliance 27–28, 32; criticisms 29; fishing communities and climate adaption 32–33, **33**
Swathi Lekshmi, P.S. 69

Tamil Nadu: 2004 Tsunami disaster 12, 40; 2004 Tsunami disaster, relief schemes 39, 41, 122; climate action plans, inadequacies 176; climate change vulnerabilities 8, **38**; climate events compensation 93–95; fishermen's foreign migration 48, 62n13; fishing castes and governance 8–10, 42–43; fishing, socioeconomic factors 8; geographic location *9*; regional climate 37–38; Sangam literature insights 6–8; surukku valai fishing ban 62n8

Tamil Nadu State Climate Change Cell (TNSCCC) 176
Taylor, Marcus 27
Thomas, A. 145
Tibby, J. 65
Tol, R.S. 24
Tripathi, T. 142
Tsunami Emergency Assistance (Sector) Project (TEAP) 39

uur panchayats (fishermen councils): climate change adaption, local decisions 125–126; coastal erosion response 99, 132–133, 135; community jurisdiction 112–113, 145–146; functions and governance systems 113–117, *115*, 146; gram panchayats, mixed interactions 117–119; localised self-governance 8–9, 11; NGO cooperation 149–150, 155, 159, 169; NGOs, local conflicts 153–154, 169; political/social capital, capacity barriers 124–125, 169–170, 173–174; terminology usage 16n4; women's exclusion 147–148, 149, 154, 155–156, 160, 169–170

Viswanathan, P.K. 129
Vivekanandan, V. 11, 117

Voss, M. 21
vulnerability research 19–20, 28–29
Vulnerability-Resilience Indicator Prototype (VRIP) 36

Wang, J. 33
Warner, Koko 101–102, 126
water salinity 92–93
Wheaton, E.E. 23
Wheeler, D. 4
Williams, M.J. 147
Wisner, B. 85
Wolf, J. 21
women's self-help groups: adaptive capacity and actions 108, 124, 154–155; community responsibilities 121–122, 151–152, 156; credit networks, necessity 152–153, 154; microfinance schemes 122, 152–153; uur panchayat interference 153, 154–155, 156–157
Woolcock, M. 25
World Bank 39, 128, 165

Yin, R. 41, 56
Yohe, G. 24

Zacharia, P.U. 74